Ion-Exchange
Chromatography
of Proteins

CHROMATOGRAPHIC SCIENCE

A Series of Monographs

Editor: JACK CAZES
Sanki Laboratories, Inc.
Sharon Hill, Pennsylvania

Ion-Exchange Chromatography of Proteins

Shuichi Yamamoto

Yamaguchi University
Tokiwadai, Ube, Japan

Kazuhiro Nakanishi

Okayama University
Tsushima, Okayama, Japan

Ryuichi Matsuno

Kyoto University
Sakyoku, Kyoto, Japan

CRC Press
Taylor & Francis Group
Boca Raton   London   New York

CRC Press is an imprint of the
Taylor & Francis Group, an **informa** business

First published 1988 by Marcel Dekker, Inc.

Published 2019 by CRC Press
Taylor & Francis Group
6000 Broken Sound Parkway NW, Suite 300
Boca Raton, FL 33487-2742

© 1988 by Taylor & Francis Group, LLC
CRC Press is an imprint of Taylor & Francis Group, an Informa business

First issued in paperback 2019

No claim to original U.S. Government works

ISBN 13: 978-0-367-45131-8 (pbk)
ISBN 13: 978-0-8247-7903-0 (hbk)

**Visit the Taylor & Francis Web site at
http://www.taylorandfrancis.com**

**and the CRC Press Web site at
http://www.crcpress.com**

Library of Congress Cataloging-in-Publication Data

Yamamoto, Shuichi
    Ion-exchange chromatography of proteins.

    (Chromatographic science ; v. 43)
    Bibliography: p. 377
    Includes index.
    1. Proteins--Analysis. 2. Ion-exchange chromatography.
I. Nakanishi, Kazuhiro. II. Matsuno, Ryuichi. III. Title.
IV. Series.
QD431.Y36 1988   547.7'5046   88.3539
ISBN 0-8247-7903-7

Preface

Ion-exchange chromatography (IEC) has been widely employed
for the separation and purification of enzymes and proteins.
In addition, recent advances in biotechnology require a large
process-scale efficient method for the separation and purification
of proteins. For this purpose, process-scale IEC will be employed
more frequently. However, the separation mechanism in IEC of
proteins is a rather complicated process. Therefore, it is not
easy to make a proper choice of the design of the column dimension and the chromatographic conditions, unless the fundamental
knowledge of this method is available.

In this book, we describe the separation mechanism of proteins
in IEC by stressing the unique characteristics of IEC of proteins.
Both theoretical and experimental works concerning this method
have been reviewed. We have also briefly described the experimental method and the apparatus commonly employed, and some
special operational procedures and apparatus. Particular empha-
sis is placed on the design and operation of the large-scale IEC.
This will be informative to the reader involved in biotechnology.

Although we have tried to emphasize physical significance
rather than mathematical manipulations, the mathematics involved
in the theoretical parts of this book may be unfamiliar to some
readers. We have attempted to guide the reader toward an
understanding of the basic concepts, and to help him to acquire
or improve the technique of IEC of proteins without calling for
a deep understanding of the mathematics. Therefore, the reader
may skip such mathematical sections. But we hope that the

reader understands the physical significance and meaning involved in the equations. This will help in understanding the experimental results.

As described in this book, IEC as a method of separating and purifying proteins is a relatively new technique and is undergoing continuous, rapid development. The literature on this method is expanding, and we have tried to include as many important references as possible. We also recognize that the scope of this book is limited to the methods that we wish to recommend and the experimental results that we found important, and that these choices are based on our backgrounds and personal preferences. So, if there are readers who feel that certain subjects deserve further exploration in a future volume, we would be grateful to have their suggestions.

Shuichi Yamamoto
Kazuhiro Nakanishi
Ryuichi Matsuno

Contents

Contents

Ion-Exchange Chromatography of Proteins

1
Introduction

An integral part of the study of biochemical science is devoted
to the separation and analysis of a particular substance. Therefore, if we look behind revolutions in life science, we find highly
developed separation techniques. For instance, in the field of
genetic engineering, which has been progressing very rapidly
for the last decade, the first step is the isolation of genes. Although there are several different methods for this step, each
method is supported by various types of effective and sophisti-
cated separation techniques. If the basic science thus established is directed to the production of a useful compound, the
productivity may be sometimes expected to be more than a hun-
dred times or more compared with that using conventional methods. However, the overall productivity of the production process
cannot be increased to as high a level as we expect from conventional production processes, and often industrialization of the
new process is abandoned. This may be partly due to the high

costs of energy and raw materials, since for the same amount of final product, the minimum energy and raw materials necessary for production are the same. However, the separation steps that follow the production step are often critical to the overall efficiency. The separation step is called a downstream treatment whose importance has been gradually recognized in the practical application of biotechnology.

Thus, we clearly see the importance of separation processes in both basic science and engineering. We should also realize that separation in engineering is much more difficult than that in basic science since the energy efficiency must be taken into consideration.

We constantly seek novel ideas of separation based on new principles. The techniques of the separation methods available at present must be improved. Finally, every separation technique should be systematized so that anyone can use it without detailed knowledge or much experience.

It is common that a theory is introduced in the systematization process, but this theory is sometimes so difficult that many users who are unfamiliar with such theoretical treatment often hesitate to be involved in such difficulties. Therefore, the theory must be as simple as possible. However, one also should recognize that theory greatly reduces the number of trial-and-error experiments that must be done to find the optimal operational conditions. In a theory, a large number of variables are often grouped with a small number of nondimensional variables. If the number of the original variables and of the grouped nondimensional variables are n and m, respectively ($m < n$), the number of experiments can be reduced by a factor of k^{m-n}, where k is the number of experiments per one variable or non-dimensional variable.

Ion-exchange chromatography (IEC) is known to be an efficient technique for the separation of proteins and has been routinely used in the biochemical science laboratory. Considering the separation efficiency and energy requirements, its relatively large-scale application to industrial protein separation is expected. In this book, we summarize and make clear the developments in the study of the ion-exchange chromatography of proteins and describe how we might design the ion-exchange chromatography of proteins with the knowledge available at present. We also discuss its inherent limitations. We can thus speculate about future problems, and these may be effective for the future systematization of protein IEC.

A large number of references on protein IEC and related subjects are published every year. A review that covers the literature on various types of chromatography is given in Analytical Chemistry every 2 years. Titles on chromatography and related methods are compiled in the bibliography section of the Journal of Chromatography in each volume. These will help the reader to search for the important references. Symposium volumes in the Journal of Chromatography are also useful for surveying current topics in the liquid chromatography of proteins. In addition to the books on (high-performance) liquid chromatography (for example, Snyder and Kirkland, 1974; Heftmann, 1975; Hearn et al., 1983), the books on the downstream treatments in biotechnology (Bailey and Ollis, 1986) and purification methods in enzymology (Scopes, 1982) are informative.

1.1 TERMINOLOGY

Before beginning, we briefly describe several terms that frequently appear in this book so that readers can easily under-
stand the text.

Figure 1.1 schematically illustrates the general operation procedure in column chromatography. Packing materials (chromatographic media) are usually packed into a cylindrical tube. Since
most commercial packing materials (the ion exchanger) for ion-
exchange chromatography of proteins have a gel structure, we
occasionally use the term "gel" for the packing material. The
term "column" usually implies the bed inside the tube. The
word column in this context thus has the same meaning as "gel
bed" or "fixed bed." The tube itself is called a column tube or
an empty column. After a sample is applied to the column, then

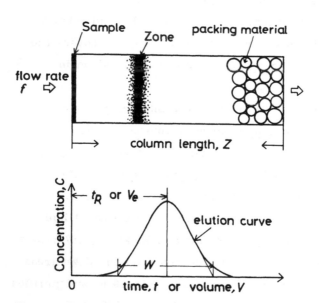

Fig. 1.1 Schematic illustration of a chromatographic column
and an elution curve.

the elution is started. The term "development" is often used
instead of "elution." However, we prefer to use "elution." The
applied sample occupies a certain part of the column. This part,
called the zone or band, gradually becomes wider (owing to
several factors described in the text) as it travels down through
the column. This phenomenon is called zone spreading, band
broadening, or peak dispersion, which is one of the important
subjects treated here.

During the elution process, the effluent (eluate) from the
outlet of the column is monitored either continuously or dis-
continuously while solvent (eluant) is fed to the inlet of the
column. Instead of the term "eluant," we very frequently use
"elution buffer." When the composition of the elution buffer is
equal to that of the inital buffer with which the column is
equilibrated, this is "isocratic elution." In contrast, if the
elution buffer is different from the initial buffer, this elution
method is "nonisocratic elution." Gradient elution and stepwise
elution are the two most common nonisocratic elution methods.
If the concentration of the sample at the outlet of the column is
plotted against the time or the volume from the start of the elution, the curve shown in Fig. 1.1 is obtained. This curve is
referred to as the "elution curve," although several different
terms, such as elution profile, elution pattern, and chromato-
gram, are also frequently used in the literature. We sometimes
call the elution curve simply the peak. The position at which
the elution curve is maximum is the peak position; the time and
the volume from the start of elution to the peak position are the
peak retention time t_R and the elution volume V_e, respectively.
Characterization of the elution curve is discussed in Sec. 2.1.1
in detail.

Generally, a chromatographic column consists of two phases: a mobile and a stationary phase (see Fig. 1.1). The mobile phase is the space occupied by the solvent (interparticle space), and its total volume in the column is the void volume V_0. The concentration in the mobile phase is given by C. There are two theoretical treatments for describing the stationary phase. In one case, the stationary phase is subdivided into several different phases. Another treatment regards the stationary phase as homogeneous. We adopt the latter since it is difficult to distinguish the phases inside ion exchangers for protein separation. The volume and concentration of the stationary phase are represented by $V_t - V_0$ and $\overline{C}$, respectively (where V_t is the total column volume). Since most ion exchangers for protein separation are of the gel type, as mentioned, we occasionally refer to the stationary phase as the gel phase.

As mentioned in the following chapters, the distribution coefficient K of a solute between two phases is a very important parameter in chromatography. However, many different definitions for K are employed in the literature. In addition, symbols employed for K and related parameters are also different. A full mathematical description of these parameters and the relations between the parameters derived, with different definitions, is presented in Sec. 2.1. However, since Sec. 2.1 presents a number of mathematical expressions, a brief description is provided here for readers who are not familar with such mathematical expressions.

In principle, the concentration has a unit of weight per volume and the distribution coefficient K between two phases is expressed as a dimensionless parameter defined by the following simple relation: $K = \overline{C}/C$. In the literature of gel filtration

chromatography, this K is sometimes designated K_{av} (Determann
and Brewer, 1975). On the other hand, the capacity factor (or
the partition ratio) k' is frequently employed in the literature of
high-performance liquid chromatography (for example, Snyder
and Kirkland, 1974). k' is equal to HK, where $H = (V_t - V_0)/$
$V_0 = (1 - \varepsilon)/\varepsilon$. ε is the void fraction of the column and can
be obtained by V_0/V_t. H implies the ratio of the volume of the
stationary phase to that of the mobile phase.

   A height equivalent to a theoretical plate (HETP) is a generally accepted measure of the chromatographic column efficiency.
HETP is defined as $HETP = Z(W/t_R)^2/16$ when the elution curve
is gaussian (see Sec. 2.1 for details). Z is the column length,
and W is the peak width at the baseline (see Fig. 1.1). The
narrower the peak, the smaller the HETP and the higher the
column efficiency is. On the other hand, as a measure of the
degree of resolution of two adjacent peaks, resolution R_s, defined as the distance between two peak maxima divided by the
sum of the half of two W values, is employed. When the elution
curves are gaussian and the area of the two peaks is the same,
R_s must be larger than 1.2 if we desire a fine separation.

   As described in the introductory section, the dimensionless
parameters are very useful and are often used in this book.
Among the various dimensionless parameters employed in this
book, we explain several important ones for readers who are not
familiar with such parameters. Instead of the time or volume
from the start of elution, we use the dimensionless parameter θ,
which is equal to V/V_0 or $t/(Z/u)$. u is the linear mobile-phase
velocity and is related to the volumetric flow rate F as $u = FZ/$
V_0. So, $\theta = 1$ is the void volume or the residence time of a
solute that can neither enter nor interact with the stationary

phase, and $\theta = 1 + H$ implies the column volume or the residence
time of a solute whose distribution coefficient $K = 1$. A void
fraction of the column usually ranges between 0.35 and 0.45.
Thus, when the void fraction is assumed to be 0.4, the column
volume is expressed by θ as $\theta = 2.5$. The term C/C_0 is the
ratio of the concentration of a solute at the outlet of the column
to the initial concentration. When $C/C_0 > 1$, the solute is concentrated. The opposite is true for $C/C_0 < 1$. X_0 is a dimensionless parameter describing the sample volume and is defined
as sample volume/V_0 or sample injection time/(Z/u). Many
other important terms and parameters appear in this book. They
are explained in the text when they appear first. Appendix G,
a list of the symbols used in this book, may also help readers.

1.2 BASIC CONCEPTS OF COLUMN CHROMATOGRAPHY

Various separation systems are classified in Table 1.1 on the
basis of the principle of separation and the operational method.
Each separation system operates according to two principles:
equilibrium separation and velocity or migration velocity separation. Batch and continuous operations are available for each
system. Pot still distillation is a typical example of equilibrium
batch separation, and mixer-settler extraction is an example of
equilibrium continuous separation. At equilibrium, the gradient
of chemical potential of components existing in the system diminishes throughout the system.

Consequently, a substance distributes between two phases
in gas-liquid, liquid-liquid, solid-liquid, gas-solid, and membrane systems. The distribution of the substance is caused
by differences in the strength of physical, chemical, and com-

bined interaction forces between the substance and each phase.
The difference in the distribution pattern of each component
enables us to separate each component. On the other hand, a
component distributes in a narrow zone in the homogeneous system in balance between the diffusional force and the external
force acting on the component in an electrophoretic system and
a centrifugal force system. The zone of each component thus
distributes in different positions along the direction of force.

During velocity separation, the separation proceeds in a nonequilibrium state in which there is a gradient of chemical potential in the system. In the electrophoretic and centrifugal force
systems, each component migrates along the direction of external
force with a different velocity caused by the external force
acting on each component. Each component is then separated
by the difference in the migration velocity. The difference in
the migration velocity in a two-phase system arises from the
difference in the strength of the interaction force of each component with two phases. Imagine a two-phase system that consists of a mobile and a stationary phase in which a number of
components move with the mobile phase in one direction. If a
component interacts with the stationary phase by attractive
force, the component diffuses into the stationary phase owing to
the local gradient of chemical potential of the component and
diffuses out again into the mobile phase when the concentration
of the component decreases. Consequently, the average migration velocity of the component becomes lower than that of the
mobile phase. Interaction forces between a component and the
stationary phase differ from component to component. The
stronger the interaction force, the lower is the migration velocity

Table 1.1 Classification of Separation Techniques

System	Principle of separation	Operation	Example
Gas-liquid	Equilibrium	Batch	Simple distillation
		Continuous	Flush distillation
	Migration velocity	Batch, continuous	Distillation tower
Liquid-liquid	Equilibrium	Batch	Batch extraction
		Continuous	Mixer-settler
	Migration velocity	Batch	Chromatography by solvent soaked in solid
		Continuous	Countercurrent extraction
Solid-liquid (solid-gas)	Equilibrium	Batch, continuous	Adsorption, desorption

		Batch	Pulse column chromatography
	Migration velocity	Continuous	Cross-flow chromatography (Fox et al., 1969)
		Continuous	Continuous chromatography by simulated moving bed (Broughton, 1968)
Electrophoresis	Equilibrium	Batch	Isoelectric point focusing electrophoresis
	Migration velocity	Batch	Disk electrophoresis
		Continuous	Cross-flow electrophoresis
Centrifugation	Equilibrium	Batch	Density gradient ultracentrifugation
	Migration velocity	Batch	Separation by sedimentation velocity method
		Continuous	Cross-flow continuous centrifugation
Membrane	Equilibrium	Batch	Dialysis
	Migration velocity	Batch	Membrane filtration
		Continuous	Cross-flow continuous filtration

of the component in the mobile phase. This is the principle of velocity separation in a two-phase system.

Velocity separation is also operated either batchwise or continuously. The sedimentation velocity method in centrifugal force, gel electrophoresis, and chromatography belongs to the batch operation. Two types of continuous operation are available. In the first type, the continuous operation is carried out by introducing a flow perpendicular to the direction of migration of the component. We can designate this as the cross-flow operation; a typical example is continuous centrifugation. The second type is most efficient, in which the two phases contact countercurrently and the components in the sample mixture are separated into one of the two phases. Countercurrent extraction and the distillation column belong to this type. Every part of the apparatus participates in the separation simultaneously, as opposed to batch and cross-flow continuous operations, in which only part of the apparatus is utilized for the separation at a time.

Liquid column chromatography, including gel filtration chromatography (GFC), hydrophobic interaction chromatography (HIC), reversed-phase chromatography (RPC), and ion-exchange chromatography, involves batch velocity separation in the liquid-solid system. Special methods for cross-flow and countercurrent continuous operation are devised for chromatography operated by the isocratic elution method, described later. Cross-flow operation is referred to as continuous chromatography (Fox et al., 1969), and countercurrent operation is carried out in a simulated moving bed (Broughton, 1968).

However, in the ion-exchange chromatography of proteins, only a batch operation is adopted. In the liquid-solid system, separation of the two phases is easy, but movement of the solid phase is difficult. This is one reason that liquid column chromatography is usually operated batchwise. This is reflected in the scale-up of the apparatus. It is difficult to scale liquid column chromatography to the capacity of distillation columns and countercurrent extractors, although the scale-up is much easier than that for separation with the centrifugal force and electrophoretic systems. Therefore we can say that liquid column chromatography has a middle-scale separation capacity.

Let us consider isocratic elution in which, for the sake of simplicity, a sample is applied as a pulse. (See also Sec. 1.1 for the meaning of the variables appearing in the following equations.) The volume of the elution buffer applied to the column while the sample pulse travels from the top to the outlet of the column is designated the elution volume. The sample travels through the void volume of the column and through part of the gel volume $V_t - V_0$, where V_t and V_0 are the total column volume and the column void volume, respectively. The elution volume V_e is thus described by Eq. (1.1):

$$V_e = V_0 + K(V_t - V_0) \qquad (1.1)$$

where K is the distribution coefficient and implies the fraction of gel volume through which the sample travels. K is also related to the sample concentration in the stationary phase $\overline{C}$ and that in the mobile phase C at equilibrium as

$$\overline{C} = KC \qquad (1.2)$$

By introducing the volume of the stationary phase to that of the mobile phase, $H = (V_t - V_0)/V_0 = (1 - \varepsilon)/\varepsilon$, Eq. (1.1) is rewritten as Eq. (1.3):

$$V_e = V_0(1 + HK) \tag{1.3}$$

where ε is the void fraction of the column.

The peak position t_R, the time necessary for the sample to travel through the column, is calculated as

$$t_R = \frac{V_e}{u_0 A_c} \tag{1.4}$$

where u_0 is the superficial velocity of the elution buffer and A_c is the cross-sectional area of the column. The migration velocity of the sample peak dz_p/dt is thus written as Eq. (1.5):

$$\frac{dz_p}{dt} = \frac{Z}{t_R} = \frac{Z u_0 A_c}{V_e} \tag{1.5}$$

By substituting Eq. (1.3) for V_e, we obtain

$$\frac{dz_p}{dt} = \frac{Z u_0 A_c}{V_0(1 + HK)} = \frac{u}{1 + HK} \tag{1.6}$$

where $u = u_0/\varepsilon$ is the linear mobile-phase velocity in the column. Equation (1.6) tells us that the migration velocity decreases with an increase in the distribution coefficient K. K values differ from component to component, reflecting the interactive force between each component and the stationary phases. This is the underlying principle of the separation and one of the most important factors affecting separation in column liquid chromatography.

Another important factor is the peak width at the outlet of the column. The narrower the peak width, the higher is the separation efficiency. As described in the following chapters, broadening of the peak width is closely related to the diffusion velocity of the sample through the laminar film around the gel

and that inside the gel and the longitudinal dispersion of the
sample in the column. For efficient operation of chromatography,
these two aspects must be clearly understood.

1.3 BRIEF HISTORY OF ION-EXCHANGE
CHROMATOGRAPHY

When the history of ion-exchange chromatography is described
(Helfferich, 1962; Heftmann, 1975), one must refer to the
success in preparing drinking water by Moses and the work of
Aristotle. As shown in Table 1.2, the use of natural or slightly
modified ion exchangers for the separation of inorganic charged
substances was promoted after the report by Thompson (1850)
and Way (1850) until phenol-formaldehyde resin was synthesized
by Adams and Holmes (1935).

The synthetic styrene and acrylic ion exchangers then played
a principal role in ion-exchange chromatography, and it became
one of the most popular separation techniques. In 1949, the
technique was applied to the separation of biochemical low-molecular-weight substances (Cohn, 1949), and in 1951, it was
incorporated into a highly developed automatic amino acid analyz-
er (Moore and Stein, 1951). In the same year, ribonuclease, a
small enzyme protein, was separated on Amberlite IRC-50 (Hirs
et al., 1951). However, such synthetic ion-exchange resins
were not used often for protein separation because of two
properties of the resin: its small pore size and its hydrophobic
nature (Fasold, 1975; Janson and Hedman, 1982). The small
pore size restricts the entrance of a protein of large size and
causes of high diffusional resistance. Proteins have a hydrophilic nature at the surface and strong hydrophobicity in the
interior. The strong hydrophobic interaction between the resin

Table 1.2 Brief History of Ion-Exchange Chromatography

Year	Researcher	Ion exchanger and use	Theoretical treatment
1850	Thompson; Way	Natural ion exchanger in soil (modified sulfonated coal)	
1855	Fick		Fick's first law for diffusion
1888	Nernst (1888; 1889); Planck (1890)		Nernst-Planck equation
1911	Donnan		Donnan equilibria
1924	Lewis and Whitman		Double-film theory
1935	Adams and Holms	Synthetic phenol-formaldehyde resin (styrene and acrylic ion exchanger)	
1940	Wilson		Equilibrium theory
1941	Martin and Synge		Plate theory (continuous)
1943	De Vault; Weiss		Nonlinear equilibrium theory
1947	Mayer and Tompkins		Plate theory (discontinuous)
1948	Gregor		Model of the ion exchanger

Year	Author		
1949	Cohn	Application of ion-exchange chromatography to separation of biologic low-molecular-weight substance	
	Glueckauf et al.		Axial dispersion coefficient
1951	Moore and Stein	Automatic amino acid analysis	
	Hirs et al.	Application of ion-exchange chromatography to protein separation (ribonuclease-Amberlite IRC-50)	
1952	Lapidus and Amundson		Mass balance model in packed bed column
1955	Drake; Freiling		Equilibrium theory for gradient elution
	Giddings and Eyring		Stochastic theory
1956	Peterson and Sober	Cellulose ion exchanger	
1956	van Deemter et al.		Relation between rate theory and plate theory
1958	Giddings		Random walk theory

Table 1.2 (Continued)

Year	Researcher	Ion exchanger and use	Theoretical treatment
1959	Porath and Flodin	Cross-linked dextran gel	
	Helfferich (1962)		Theoretical treatment of equilibrium and kinetics in <u>Ion Exchanger</u>
1961	Hjerten	Polyacrylamide gel	
	Hjerten and Mosbach	Agarose gel	
1965	Kubin; Kucera		Application of mass balance model and moment method to chromatography
1971	Porath et al.	Cross-linked agarose gel	
1976	Chang et al.	Silica-based high-performance ion-exchange packing materials	
	Pitt		Elution behavior in ion-exchange chromatography of small molecules
1982	Kato et al.	Hydrophilic porous vinyl polymer gel	

and a protein makes the elution of the protein after adsorption very difficult and sometimes causes denaturation of the protein or enzyme. An ion exchanger with a large pore size and hydrophilic nature was sought. The cellulose ion exchangers developed by Peterson and Sober in 1956 were the first hydrophilic ion exchangers. In addition to this development, the successive synthesis of cross-linked dextran ion-exchange gels in the 1960s after the synthesis of cross-linked dextran gels for gel filtration chromatography in 1959 by Porath and Flodin firmly established ion-exchange chromatography as a separation technique for proteins. It is no exaggeration to say that we cannot proceed with work related to proteins without ion-exchange chromatography.

In spite of the excellent properties of these hydrophilic gels, softness and swelling or shrinking properties impeded good separation. Developments directed at the synthesis of hard hydrophilic ion exchangers resulted in the cross-linked agarose ion-exchange gels (Porath et al., 1971), hydrophilic vinyl polymer ion-exchange gels (Kato et al., 1982a), and others. Ion-exchange packing materials of very small particle size (about 10 μm diameter), such as ion-exchange silica gels (Chang et al., 1976) and ion-exchange hydrophilic polymer gels (Kato et al., 1983b), have enabled high-performance IEC (HPIEC) of proteins. These new types of ion-exchange packing materials give a high separation efficiency and a high speed of separation and will be used more frequently for preparative separation as well as analytical separation.

Theoretical developments related to column chromatography are also summarized in Table 1.2. Discovery of Fick's first law for diffusion (Fick, 1855), film theory (Lewis and Whitman, 1924),

and the concept of axial dispersion (Glueckauf et al., 1949) reflected on the mass balance model (Lapidus and Amundson, 1952),
which rigorously describes the phenomena occurring in column
chromatography. Plate theory (Martin and Synge, 1941) is a
simple model that reduces the mathematical difficulty of analyzing
chromatographic behavior, and the random walk model proposed
by Giddings (1958) was based on the stochastic theory (Giddings
and Eyring, 1955). These as well as the equilibrium theory
(Wilson, 1940; DeVault, 1943; Weiss, 1943; Drake, 1955) are the
distinguishable models common to any type of chromatography.
The work of van Deemter et al. (1956) and the application of a
moment method to chromatography by Kubin (1965) and Kucera
(1965) played important roles in the progress of chromatographic
theory. An important theory directly related to the ion exchanger is the Donnan membrane equilibrium (Donnan, 1911) and
the Nernst-Planck equation (Nernst, 1888; 1889; Planck, 1880).
A model of the ion exchanger was first proposed by Gregor in
1948. Theoretical works on equilibrium and kinetics of the ion
exchanger were summarized by Helfferich in 1959, including
his own excellent works.

1.4 COMPARISON OF ION-EXCHANGE CHROMATOGRAPHY WITH OTHER CHROMATOGRAPHIC TECHNIQUES

As mentioned in the previous section, chromatographic separation
is based on differences in the migration velocity governed by the
distribution coefficient K of components in the sample mixture.
The distribution of a component between the mobile and stationary phases occurs owing to various types of noncovalent binding interactions, such as hydrogen bonding, hydrophobic interaction, electrostatic interaction, and van der Waals interaction

and also to a size-exclusion (molecular sieving) effect. The dependence of K on particular variables not only characterizes the nature of a given type of chromatography but also determines the suitable elution method. We survey several types of column liquid chromatography commonly employed for protein separation, which are briefly summarized in Table 1.3.

The principle of separation by gel filtration chromatography, often called gel permeation chromatography (GPC) or size-exclusion chromatography (SEC), is based on the size-exclusion effect. The distribution coefficient K in GFC is therefore dependent only on the size of molecules. For smaller molecules, which can freely diffuse into the pores of the gel, K is nearly equal to 1, but K is zero for very large molecules since they cannot enter the pores. Since K is independent of the sample concentration and the composition of an elution buffer under the usual conditions, GFC is operated isocratically. Consequently, all substances contained in the sample are eluted within one column volume in order of decreasing molecular size. This simple elution mechanism is an attractive feature of GFC, although it also limits the maximum number of resolvable components (Giddings, 1967). Since the elution is isocratic, proteins are not subjected to severe conditions, such as high salt and acidic or alkaline solutions. This is also one advantage of GFC. We cannot expect the concentration effect, however, and therefore the sample volume is a major factor limiting separation efficiency.

Hydrophobic interaction between a substance and a constituent of gel particles causes the distribution in hydrophobic interaction chromatography. The distribution coefficient K depends on ionic strength I and on the pH in the surrounding solution,

Table 1.3 Classification of Liquid Chromatography of Proteins

Type	Elution method	Distribution coefficient[a]	Separation principle	Moving rate of zone
Gel filtration	Isocratic	Constant	Molecular shape	Constant
Ion exchange	Gradient	$f(I, pH, C)$	Electrical charge	Variable
Hydrophobic	or	$f(I, pH, C)$	Hydrophobicity	Variable
Affinity	stepwise	$f(I, pH, C, C_e)$	Biospecific adsorption	Variable

[a]The distribution coefficient also depends on the structure of the matrix of packing materials and on the shape of proteins; I, ionic strength, C, concentration of protein; C_e, concentration of a component with a specific affinity of proteins or enzymes.

which often exceeds 1. Therefore, nonisocratic elution, in
which the ionic strength is decreased, is preferred.

The separation principle of reversed-phase chromatography
(RPC) is essentially the same as that of HIC. Although proteins
are separated on the basis of their hydrophobicity in HIC and
RPC, the alkyl chain of ligand in RPC is longer and the ligand
concentration in RPC is higher than in HIC. Consequently,
proteins are adsorbed strongly on RPC columns owing to the
high hydrophobicity of RPC packings. The desorption of proteins from the RPC column must be performed with organic
solvents. Since this denatures most proteins, RPC has been
employed as an analytical method rather than as a preparative
method for the separation of proteins (Regnier and Gooding,
1980; Krstulovic and Brown, 1982). However, since the high
separation efficiency of RPC is very attractive, strategies for
recovering the biologic activities of proteins by RPC have been
recently developed (Hearn, 1984). RPC may thus be recognized
as a future preparative separation method for proteins.

In our objective, ion-exchange chromatography, the electrostatic repulsion and attraction forces of a gel with a large number of charged groups against a charged solute molecule dis-
tribute that molecule. K is usually a function of I, pH, and
protein concentration and ranges from zero to more than one.
In some cases, K exceeds 10^5. This wide range of K values is
one of the characteristics of IEC that makes this method attractive, as shown later.

In these three chromatographic methods, each, in principle,
incorporates only one kind of interaction, but affinity chromatography exerts a concerted action of many kinds of interactions. Consequently, an affinity gel with a special ligand

distinguishes a species of a molecule and specifically binds the molecule by means of a strong interaction. Thus, one may expect affinity chromatography to be a highly efficient tool to separate and purify molecule in one step from a mixture containing many kinds of molecules with closely similar physical properties.

Although the specificity of IEC to a substance is not as strong as that of affinity chromatography, one may expect a fine separation owing to the versatility of the charge state of the substance, such as the kind, polarity, and number of charged groups. Moreover, gel particles with a high capacity are easily prepared and the operational method is simple. Because of these advantages, IEC is now an essential tool for the separation of biochemical substances.

We examine briefly how the separation in IEC is affected by various factors on the basis of the thermodynamic concept. At equilibrium, the chemical potential of a molecular species is considered identical in both the stationary (gel) and the mobile phases. The chemical potential of a species μ_i is written as

$$\mu_i = \mu_{i0} + \mathscr{R} T \ln a_i + Z_i \mathscr{F} \Delta \phi + v_i \Delta P \tag{1.7}$$

where

$\quad \mu_{i0}$ = chemical potential in the standard state

$\quad a_i \quad$ = activity

$\quad \mathscr{R} \quad$ = gas constant

$\quad T \quad$ = absolute temperature

$\quad Z_i \quad$ = number of the charge with polarity

$\quad \mathscr{F} \quad$ = Faraday constant

$\quad \phi \quad$ = electrostatic potential

$\quad v_i \quad$ = specific volume

$\quad P \quad$ = pressure

The second term in Eq. (1.7) denotes the concentration term,
the third term, the electrostatic potential term, and the fourth,
the pressure term. The distribution of a species in ion-exchange
chromatography depends mainly on the concentration and electrostatic potential terms. However, when a gel is used that has a
high degree of cross-linking the pressure term should be considered because of the high swelling pressure, especially for a
high-molecular-weight species with a large specific volume.
Many other factors affect the distribution of a species. The
kind, number, and pK values of charged groups are important
properties of the molecule, and the kind, pK value, and concentration of charged groups attached to the gel are the important properties of the gel.

Ion pair formation between the charged groups of molecule
and an ion-exchange gel is another important factor. Thus, the
resultant distribution coefficient K of a given substance is a
function of the operational conditions, the pH and the ionic
strength of the surrounding solution, and the concentration of
the substance. In the usual batch operation procedure of IEC,
a sample mixture dissolved in an appropriate buffer is applied
at the top of the column equilibrated with that buffer. This
buffer is referred to here as the initial starting buffer. An
elution buffer is then introduced to elute the sample adsorbed
at the top of the column. When the distribution coefficient K
in the initial starting buffer is so high that no migration of the
sample is expected, we must change the composition of the elution buffer to reduce K. This procedure is called nonisocratic
elution. When the initial and elution buffers are identical, the
elution is identified as isocratic. To reduce the distribution coefficient, the ionic strength of the elution buffer is increased

or the pH is shifted near the isoelectric point of the objective
substance. When we theoretically treat elution behavior in ion-
exchange chromatography, we must consider that at least two
components are exchanged between the gel and the mobile
phases. When the pH change during the elution is significant,
at least three components are involved in the exchange process.
The number of differential equations that describe the elution
behavior increases in proportion to the number of components
to be considered. An increase in the number of differential
equations accelerates the difficulty of the mathematics. Another
difficulty with the theoretical treatments is that the components
concerned are ion species. With ions, we cannot adopt the
simple Fick's law for diffusion but must use the Nernst-Planck
equation (Nernst, 1888; Planck, 1890), shown in Eq. (1.8).

$$N_i = -D_i \left(\text{grad } C_i + Z_i C_i \frac{\mathscr{F}}{\mathscr{R}T} \text{ grad } \phi \right) \tag{1.8}$$

where N_i = flux of ion species i and D_i = diffusion coefficient of
ion i. The second term on the right-hand side of Eq. (1.8)
represents the mass flux caused by the gradient of the electrostatic field. Although no electrostatic field is applied to the
column, the gradient of electrostatic potential is induced owing
to the difference in the diffusional velocity of each ion species.
Let us consider the situation in which a cation A and an anion
B diffuse in the same direction. The diffusion coefficient of A
is greater than that of B. If A diffuses faster than B, an electric current is induced. However, since there is no electron
acceptor in the system, the electric current should not appear.
To satisfy this situation, an electrostatic force must be induced
that reduces the diffusion velocity of A and enhances that of

B. Helfferich (1962) derived an equation for the apparent diffusion coefficient of an ion by applying the Nernst=Planck equation to diffusion phenomena, assuming no electric current and
electric neutrality in the ion exchanger. The resultant apparent
diffusion coefficient of the two ion species system is a function
of the diffusion coefficients, the charge numbers, and the concentrations of the two ion species. This means that the apparent diffusion coefficient changes with time and is not constant along the axial direction of the column or the radial direction in the exchanger.

1.5 CHARACTERISTICS OF ION-EXCHANGE
CHROMATOGRAPHY OF PROTEINS

Proteins are high-molecular-weight compounds of amino acids
linearly polymerized by peptide bonds. Therefore, they possess
a large number of dissociable groups, that is, terminal amino
and carboxyl groups, carboxyl groups of asparatyl and glutamyl
side chains, an amino group with a lysine side chain, a guani-
dinium group with an arginine side chain, and an imidazole group
contained in histidine. For instance, ribonuclease (RNase) has
34 dissociable groups with its isoelectric point pI at pH 9.6
(Lehninger, 1970). Such dissociable groups are distributed on
the surface in globular proteins. The pK values of these groups
differ even within the same kind of group, depending on the
micromolecular environment. The net charge is zero at the isoelectric point pI. The number of net negative charges increases
with pH above the pI. Similarly, net positive charges increase
with a decrease in pH below the pI (see Fig. 1.2). Therefore,
proteins can be looked upon as polyvalent amphoteric ions.
These properties of proteins strongly affect their distribution

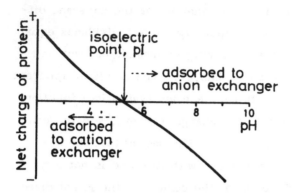

Fig. 1.2 Schematic representation of protein net charge and protein adsorption behavior to ion exchangers as a function of pH. pI is the isoelectric point of proteins.

between the mobile and the gel phases in IEC. As described in Chap. 3, the distribution coefficient K is strongly dependent on the number of net charge of the proteins. Therefore, K is sharply dependent on pH. When the number of the net charge is large, a slight change in the ionic strength results in a great change in K. For instance, in the case of β-lactoglobulin A on a commercial DEAE ion-exchange gel at pH 7.7 and 20°C (see Chap. 3), K reduces from 500 to 1 as I is increased from 0.1 to 0.3. This also means that a slight increase in I causes an abrupt acceleration in the migration of the protein zone in the column, as shown in Eq. (1.6). As mentioned in Sec. 1.4, the usual elution procedure for the IEC of proteins is noniso-cratic elution. Linear gradient elution, in which the ionic strength of the elution buffer, which is composed of low-molecular-weight salts, is increased linearly, is most commonly employed. In this elution method, the ionic strength of an initial starting buffer with which the column is equilibrated is taken as low so that the proteins are adsorbed in a thin layer at the top of the column. In this buffer, almost no migration is expected except that of proteins with low K values even at the reduced ionic strength or with charges of the same polarity as that of the ion exchanger. Such proteins move through the column in the initial starting buffer. On the other hand, the K of proteins adsorbed at the top of the column decreases to near 1 with an increase in the ionic strength I of the elution buffer. Therefore, the zones of the proteins first move slowly and then accelerate owing to the linear increase in I. Finally, they move with the same speed as that of the elution buffer. In this way, proteins in the sample mixture are separated.

Elution behavior is also affected by properties of the proteins. Because of their high molecular weights, the diffusion coefficients of proteins are far smaller than those of salts. For instance, the molecular diffusion coefficient of bovine serum albumin at $20°C$, 6.1×10^{-7} cm^2/s, is less than 1/20 that of NaCl, 1.4×10^{-5} cm^2/s. Moreover, when the ionic strength is low so that a large amount of a protein is adsorbed to the ion exchanger, the diffusion coefficient of the protein in the ion exchanger becomes much lower than the molecular diffusion coefficient (see Chap. 4). This means that the effective diffusion coefficient is extremely low compared with that of salts when the ionic strength is low. To analyze the elution behavior in ion-exchange chromatography, at least two components must be taken into consideration (see Chap. 3). The elution behavior of one component affects that of another, and this mutual interaction makes the mathematics extremely difficult. However, in the IEC of proteins, an expedient theoretical treatment may be devised by taking into account the extreme differences between the properties of the two components, salts and proteins. Since the distribution and diffusion coefficients of salts are far smaller and larger, respectively, than those of proteins, the elution of salts is carried out nearer the equilibrium state compared with the elution of proteins. Moreover, the molar concentration of salts is far greater than that of proteins under the usual conditions of protein IEC. Therefore, the elution behavior of salts may be independent of that of proteins. For the same reasons, the effect of the local electrostatic potential induced by the difference in the diffusional velocities of ion species may be ignored in the calculation of the elution behavior of proteins. However, the effects of salt concentration are incorporated into the analysis of the elution of a protein through the distribution coefficient and effective diffusion coefficient of the protein in the ion exchanger, both of which depend on the salt concentration (ionic strength). The equilibrium distribution coefficient is determined experimentally as a function of salt concentration in the mobile phase. Since we assume that the elution of a salt is performed near the equilibrium state, the distribution coefficient determined by equilibrium experiments and the resultant effective diffusion coefficient can be used for the analysis of the elution behavior of proteins. With the simplification described earlier, ion-exchange chromatography of proteins is regarded as adsorption chromatography. We use this expedient treatment for establishing the theoretical model of elution behavior, as shown in Sec. 2.2. Although this treatment is not strict, experimental elution behavior is reproduced fairly well, as shown in Chap. 8.

1.6 VARIOUS ELUTION PROCEDURES OF ION-EXCHANGE CHROMATOGRAPHY OF PROTEINS

Both isocratic and nonisocratic elution procedures are adopted for the ion-exchange chromatography of proteins (see Fig. 1.3).

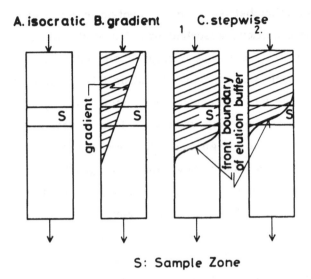

S: Sample Zone

Fig. 1.3   Schematic drawing of various elution procedures.

However, the use of the isocratic elution method for protein IEC, in which the same buffer solution is used for the equilibration of the column, the sample solution, and the elution, is restricted to special cases since the distribution coefficients of proteins in the sample solution to be separated cover a broad range. Isocratic elution is used, for example, when only the surface of the exchanger is used for the adsorption of proteins. Because of the low capacity for adsorption, the distribution coefficient is low enough to adopt isocratic elution.

Stepwise and gradient elution procedures belong to the category of nonisocratic elution in which the composition of the elution buffer is different from that of the initial buffer and is most commonly employed for protein IEC. With these procedures, proteins with distribution coefficients of broad range under the initial conditions are gradually eluted in order of the magnitude

of the force of interaction between a given protein and the exchanger. In stepwise elution, adsorbed proteins at the top of
the column are eluted by a stepwise change in pH and/or I;
in gradient elution, such a change in pH and/or I is continuous.
A linear increase in the ionic strength, designated linear gradient elution, is the most frequently used procedure in various
types of gradient elution, although many different types of
devices are available for making various shapes of gradient.

There are many discussions about the advantages and disadvantages of gradient and stepwise elution (Saunders, 1975;
Snyder and Saunders, 1969; Morris and Morris, 1964). The disadvantage of using the apparatus for making a gradient is not a
severe problem compared with the many advantages of gradient
elution. First, proteins in the sample mixture with a broad
range of K values at the ionic strength of the initial starting
buffer are eluted in a short time without using a large amount
of the elution buffer. This makes it possible to survey the
elution behavior of unknown samples very rapidly and to use
automatic operation. Because of the zone sharpening effect of
gradient elution (see Sec. 2.2), dilution of the protein concentration is reduced, or in some cases, the protein concentration
increases at the exit of the column.

On the other hand, in stepwise elution, the special case of
gradient elution, it takes a long time to complete the elution
with a large amount of buffer and automatic operation is difficult.
Although the zone sharpening effect is expected when a protein
is eluted with a spreading boundary of the elution buffer (see
C.2 in Fig. 1.3), proteins eluted with the plateau are diluted
(see C.1 in Fig. 1.3). Moreover, when more than one protein
is eluted in the same spreading boundary, separation cannot be

expected and the proteins are eluted as a single peak.   In
contrast, we often encounter a peak of one component split into
two peaks (see Sec. 8.4).   The peak eluted later is often called
a false or artificial peak.   This phenomenon appears if we
devised the wrong elution schedule for the stepwise change of
the elution buffer, especially with a protein to be separated that
obeys a nonlinear adsorption isotherm.

If we fix our attention on the resolution of peaks, a different
conclusion is reached.   As shown in Sec. 2.2.4, a measure of
peak resolution R_s, which is defined as the ratio between the
distance of two peaks to the width of the peak, is inversely
proportional to the square root of the slope of the gradient.
This means that the steeper the slope of the gradient, the lower
the resolution.   The limit of the shallowest slope of the gradient
is the stepwise increase of the elution buffer in which the peak
is eluted after the elution buffer reaches the outlet of the
column (see C.1 in Fig. 1.3).   The resolution is highest in this
type of stepwise elution method.   Therefore, as long as we
take special care that the elution schedule does not cause the
false peak and that not more than one protein is eluted with the
spreading boundary of an elution buffer, stepwise elution gives
better resolution than linear gradient elution.   The choice of
elution procedure depends on the user's conditions.   Keep in
mind that such a choice must be made after careful consideration
of the operating time, the amount and concentration of the elution buffer, the concentrations of eluted proteins, and the
resolution of peaks.

2
Theoretical Aspects

This chapter deals with theories of chromatography. We can design or operate chromatographic processes very easily if we have fundamental knowledge of the theories of chromatography. Since we do not intend to describe the theories of chromatography in detail, only the basic equations and/or the analytical solutions are given together with the basic idea and the important assumptions in the theory. For readers interested in details of each chromatographic theory, the original papers quoted in this chapter and the following excellent reviews or books and references therein give more than adequate coverage (Giddings, 1965; Grushka et al., 1975; Heftman, 1975; Morris and Morris, 1964; Yang and Tsao, 1982). In addition, some of the reviews or the textbooks for a fixed-bed adsorption process or residence time distribution will be also helpful in understanding the theories of chromatography (for example, Lightfoot et al., 1962; Sherwood et al., 1975; Vermeulen et al., 1984).

In Sec. 2.1, general theories of isocratic elution chromatography are reviewed. Therefore, when ion-exchange chromatog-

raphy is operated isocratically, these theories may be applied.
Special attention is paid to the correspondence between different
theories of linear isocratic elution chromatography.   Section 2.2
treats the theory of ion-exchange chromatography.   The elution
mechanisms of linear gradient elution and of stepwise elution are
illustrated qualitatively and then a model presented by the
authors is described (Yamamoto et al., 1983a).   Simple methods
for predicting the peak position and width in linear gradient
elution are also proposed.   A resolution equation is also derived
for the linear gradient elution.

     We do not mention the values of the parameters appearing in
the height equivalent to a theoretical plate (HETP) or N_p equation and their dependence on several important variables, such
as particle diameter and temperature.   These topics are treated
in Chaps. 4, 5, and 8.

2.1   GENERAL THEORIES OF CHROMATOGRAPHY

Although many models or theories have been proposed for chromatography, much of this section is devoted to three types of
chromatographic theory: equilibrium theory, mass balance model,
and palte theory.   The second is also called a rate theory or a
dispersion model.   The plate theory is also frequently termed a
tank-in-series model.   As pointed out by Grushka et al. (1975),
the development of chromatographic theories may be not sequen-
tial but parallel.   However, these three theories made a large
contribution to the development of chromatographic theories.
The equilibrium theory, in which zone spreading effects are
ignored, is a limiting form of the mass balance model, the most
rigorous and complicated theory, as shown later.   A concept of

HETP derived from the plate theory is now accepted as a measure of the column efficiency and is introduced in the mass
balance model. We therefore describe the three theories sequentially in Secs. 2.1.2 through 2.1.4, presenting their physical meanings qualitatively and pointing out their advantages and
disadvantages. In Sec. 2.1.5, equations for the number of
theoretical plates N_p and HETP derived from different models
are compared and discussed. The correspondence between different models is also illustrated. Several typical experimental
results for HETP are shown in Sec. 2.1.6 and examined on the
basis of the HETP equation given in Sec. 2.1.5. Although HETP
is a useful measure of column efficiency, it is not directly em-
ployable as a measure of the separation efficiency of two substances. For this purpose, resolution R_s, defined as the ratio
of the distance between the peaks of the two elution curves to
the half of the sum of their widths, is conveniently used. An
R_s equation is also derived and discussed in Sec. 2.1.7.

Most of this section is devoted to isocratic elution chromatography. Several important theoretical treatments, such as a
constant-pattern approach and frontal analysis (Vermeulen et al.,
1984), are thus not described. Frontal analysis is considered
an efficient method for the determination of the adsorption
equilibrium, as stated by Morris and Morris (1964). We determined adsorption equilibrium between proteins and ion exchangers
using frontal analysis (Yamamoto et al., 1983b). The results
are shown in Chap. 3. The mechanism of displacement chromatography is described qualitatively in Sec. 8.5. Its theoretical treatment is found in the literature (Helfferich and James,
1970; Rhee and Amundson, 1982; Frenz and Horvath, 1985).

2.1.1 Characterization of the Elution Curve
by Statistical Moments

Before describing theories of chromatography, we demonstrate
the importance of statistical moments in characterizing the shape
of the elution curve. The elution curve $C(t)$ is also denoted
the distribution curve with respect to the residence time t of a
solute in the column.

The nth normalized statistical and central moments of the
elution curve are defined by Eqs. (2.1) and (2.2), respectively:

$$\mu'_n = \frac{\int_0^\infty C(t)t^n \, dt}{\int_0^\infty C(t) \, dt} \tag{2.1}$$

$$\mu_n = \frac{\int_0^\infty C(t)(t - \mu'_1)^n \, dt}{\int_0^\infty C(t) \, dt} \tag{2.2}$$

The first statistical moment μ'_1 implies the mean residence time
of the curve; the second central moment μ_2 indicates the variance around the mean residence time. Although moments higher
than the second central moments are indicative of the magnitude
of the peak asymmetry or the peak flatness, the following two
definitions are most frequently used:

$$\text{Skewness} = \frac{\mu_3}{\mu_2^{3/2}} \tag{2.3}$$

$$\text{Kurtosis} = \frac{\mu_4}{\mu_2^2} - 3 \tag{2.4}$$

Figure 2.1 shows two examples of the elution curve. Curve
A is symmetrical and can be approximated to a gaussian error

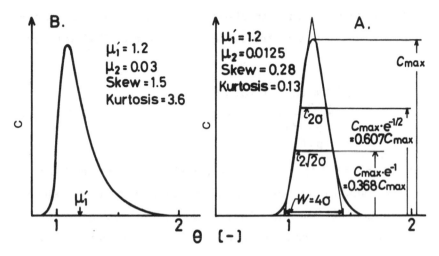

Fig. 2.1 Symmetrical and asymmetric elution curves that give the same mean retention time.

curve. As stated previously, it is desirable in designing chromatographic separation processes to predict the peak retention time t_R and the width of the curve at the baseline W of the elution curve. For a curve close to the gaussian shape, such as curve A, μ_1' is almost equal to t_R. The variance μ_2 can be approximated to the value of σ^2 calculated from the width at $C = C_{max}\, e^{-1} = 0.3679 C_{max}$ and at $C = C_{max}\, e^{-1/2} = 0.6066\, C_{max}$ or at the baseline, as shown in Fig. 2.1. The values of skewness and kurtosis approach zero. On the other hand, μ_1' is not identical to t_R in a skewed curve like curve B, shown in Fig. 2.1, although its μ_1' is the same as that of curve A. The values of skewness and kurtosis are much larger than zero. It is therefore stressed that curves with the same μ_1' and μ_2 values are not always identical. In addition to the statistical moments, several methods are available for measuring column efficiency (Bidlingmeyer and Vincent Warren, 1984).

It is known that the nth statistical moment μ'_n can be related
to the nth derivative of the Laplace transform of C(t) through
Eq. (2.5):

$$\mu'_n = \frac{(-1)^n \lim_{p \to 0} (d/dp)^n \tilde{C}(p)}{\lim_{p \to 0} \tilde{C}(p)} \tag{2.5}$$

where $\tilde{C}(p) = \mathcal{L}C(t)$, the Laplace transform of C(t), and p =
Laplace transform variable. This relation is very useful since
the equation for moments of the elution curve, which gives detailed information about the nature of the elution curve, can be
related to the operating and column variables even when the
analytical solution of the partial or ordinary differential equation is unknown. This method is called the moment method.
Application of the moment method is described in the following
sections. (For the details of the moment method, see Grushka,
1975; Kubin, 1965; Kucera, 1965; Ramachandran and Smith,
1978.)

2.1.2 Equilibrium Theory

As the name suggests, this theory is based on the assumption
that the equilibrium of the concentration of a solute between the
mobile and stationary phases is always established during the
elution process. Linear equilibrium theory was first treated by
Wilson (1940), which is the first theoretical treatment of chromatography, as Giddings (1965) stated. Weiss (1943) and De
Vault (1943) extended the equilibrium theory to nonlinear chromatography. Although the equilibrium theory is too simplified
to describe the actual chromatographic process, it does give
important information on the movement of the zone of the solute

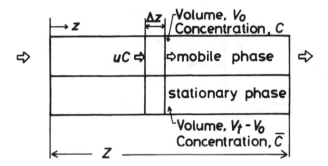

Fig. 2.2   Mass balance in a column according to the equilibrium theory.

in the column.   Therefore, we describe here several important subjects common to the following sections.

Let consider a mass balance in a small element of a column, as shown in Fig. 2.2.   (For the meanings of the symbols, see Sec. 1.1.)   We assume that the velocity profile of a flow is flat (a plug flow) and that the equilibrium of the solute concentration between the two phases is instantaneously established.   With these assumptions, a general basic equation in the equilibrium theory is given by the partial differential equation

$$\frac{\partial C}{\partial t} + H \frac{\partial \overline{C}}{\partial t} = -u \frac{\partial C}{\partial z} \tag{2.6}$$

The first and second terms on the left-hand side of Eq. (2.6) imply accumulation of the solute in the mobile and the stationary phases, respectively.   The term on the right-hand side represents mass transport by convection.

Adsorption or distribution equilibrium between C and $\overline{C}$ can be roughly grouped into three types, as shown in Fig. 2.3. One is the linear isotherm shown by curve A where the distribution coefficient $K = \overline{C}/C$ is constant.   Nonlinear isotherms can be

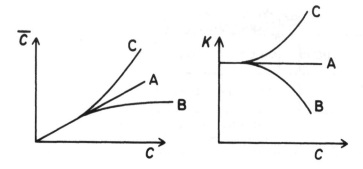

Fig. 2.3   Shapes of adsorption isotherms and distribution coefficients.

subdivided into two types: one is convex upward (curve B) and another is concave upward (curve C).   In a fixed-bed adsorption process, the former is called a favorable isotherm and the latter unfavorable (Vermeulen et al., 1984).   In other words, K is a decreasing function of C for curve B (favorable isotherm), but for curve C (unfavorable isotherm) it is an increasing function of C.   Actually, isotherms in IEC are usually favorable, expressed by, for example, the Langmuir equation and the Freundlich equation (see Chap. 3).

We first consider linear isocratic elution chromatography, where $K = \overline{C}/C$ is constant.   Then, Eq. (2.6) is rearranged as

$$(1 + HK)\ \frac{\partial C}{\partial t} = -\ \frac{u\,\partial C}{\partial z} \tag{2.7}$$

There are two different boundary conditions at the inlet of the column.   In one, a solute is fed continuously.   The boundary condition is given by

$$C = C_0 \text{ for } t \geqslant 0 \text{ at } z = 0 \tag{2.8a}$$

where C_0 is the initial concentration of the solute. This boundary condition is called a step input, and the method itself is referred to as frontal analysis. A plot of the concentration of the solute at the outlet of the column as a function of time is often designated a breakthrough curve.

Another boundary condition is given by

$$C = C_0 \quad \text{for } 0 \leqslant t \leqslant t_0 \text{ at } z = 0$$
$$C = 0 \quad \text{for } t > t_0 \text{ at } z = 0$$

(2.8b)

This implies that a solute is introduced during the sample injection time t_0 and then the elution is started. This is the boundary condition that concerns us, since it describes the usual elution method of liquid chromatography. When t_0 is small, that is, when a sample volume is small, Eq. (2.8b) is called a pulse input.

The analytical solutions of Eq. (2.7) for a step input (frontal analysis) and a pulse input (elution chromatography) are obtained as

$$C = C_0 \quad \text{for} \quad 1 + HK \leqslant \theta \leqslant 1 + HK + X_0 \quad \text{for pulse input}$$

(2.9a)

$$C = C_0 \quad \text{for} \quad \theta \geqslant 1 + HK \quad \text{for step input} \quad (2.9b)$$

where $\theta = t/(Z/u) = t/\tau$ = dimensionless time and X_0 = ratio of the sample volume to the void volume of the column or the ratio of the sample injection time to the residence time of nonretained solute; that is, $X_0 = t_0/\tau$. These two solutions tell us that the solute applied at the inlet at a given time is eluted from the column after the dimensionless time $\theta = 1 + HK$ has passed. If the input pulse is negligibly small or if we focus our attention on the center of the zone z_p, the following equation is obtained:

$$\frac{dz_p}{dt} = \frac{u}{1 + HK} \qquad (2.10)$$

It follows from this equation that the moving velocity of the zone is constant at a given HK. The peak position (peak retention time) with respect to time t_R and that to the dimensionless time θ_R are given by the equations

$$t_R = \frac{Z}{u}(1 + HK) = \tau(1 + HK) \qquad (2.11)$$

$$\theta_R = 1 + HK \qquad (2.12)$$

Let us define a variable R as a fraction of a solute presented in the mobile phase in relation to that in both phases at a certain time.

$$R = \frac{CV_0}{CV_0 + \overline{C}(V_t - V_0)} = \frac{1}{1 + HK} \qquad (2.13)$$

Insertion of Eq. (2.13) into Eqs. (2.11) and (2.12) gives

$$t_R = \frac{Z/u}{R} = \frac{\tau}{R} \qquad (2.11')$$

$$\theta_R = \frac{1}{R} \qquad (2.12')$$

From Eq. (2.10), we can interpret R as $R = (dz_p/dt)/u$ as the ratio of the moving velocity of the zone to the mobile-phase velocity, which is another definition of R. In contrast to R, we define R_f as the ratio of the moving velocity of the zone to the moving velocity of the elution buffer. R_f finds its importance in nonisocratic elution.

Next, we consider nonlinear isocratic elution. In this case, Eq. (2.6) becomes

$$\left[1 + H \frac{\partial \overline{C}}{\partial C}\right] \frac{\partial C}{\partial t} = - \frac{u \partial C}{\partial z} \qquad (2.14)$$

If we take the coordinate in which C is constant with respect to time, the total time derivative dC/dt becomes zero. Then,

$$\frac{dC}{dt} = \frac{\partial C}{\partial t} + \left(\frac{dz}{dt}\right)_c \frac{\partial C}{\partial z} = 0 \qquad (2.15)$$

From these two equations, we obtain

$$\left(\frac{dz}{dt}\right)_c = \frac{u}{1 + H(\partial \overline{C}/\partial C)_c} \qquad (2.16)$$

The general solution for the elution curve is given by

$$t = \left(\frac{Z}{u}\right) \left[1 + H \left(\frac{\partial \overline{C}}{\partial C}\right)_c\right] \qquad (2.17)$$

which is rewritten in the dimensionless form as

$$\theta = 1 + H \left(\frac{\partial \overline{C}}{\partial C}\right)_c \qquad (2.18)$$

We can see from these equations that the moving velocity of the position with a certain C is constant and is determined by the slope of the isotherm. However, it should be noted that when a solute is applied continuously and K is a decreasing function of C (favorable isotherm), the front of the zone moves with a velocity $u/(1 + HK_0)$, where K_0 is the concentration-dependent distribution coefficient at $C = C_0$. Therefore, for step input (frontal analysis), the breakthrough curve can be represented by

$$C = C_0 \quad \text{for } \theta > 1 + HK_0$$

$$C = 0 \quad \text{for } 0 < \theta \leqslant 1 + HK_0$$

$$(2.19)$$

Note that Eq. (2.19) is also important for the determination of the length of the adsorption zone during sample application in ion-exchange chromatography, as shown in Sec. 8.1.2.

When the solute applied to the column is small (pulse input) and is given by X_0, the dimensionless time for a given C can be expressed by

$$\theta = X_0 + 1 + H \left(\frac{\partial \bar{C}}{\partial C} \right)_c \qquad (2.20)$$

This equation is valid until the integrated area under the elution curve becomes equal to $C_0 X_0$. There is a case in which the elution curve retains its initial concentration C_0 when X_0 is large. Weiss (1943) termed this a "partly developed band"; another is called a "fully developed band." Various cases, including favorable and unfavorable isotherms for fully or partly developed zones, are treated by Weiss (1943), Houghton (1963), and Dunckhorst and Houghton (1966). Finally, as an example, the elution curves are calculated for the case in which K is assumed to decrease with C linearly; that is, $HK = 1 - K_1 (C/C_0)$, shown in Fig. 2.4. It is seen that the elution curves for the nonlinear isotherm have a sharp front and a long tail. In addition, the magnitude of the asymmetry and the length of the tail increase with the concentration dependence of K. This is the origin of the tail due to the nonlinear isotherm. In actual chromatographic processes, the zone spreading effects due to various factors described later broaden the sharp front boundary. However, skewed elution curves similar to those in Fig.

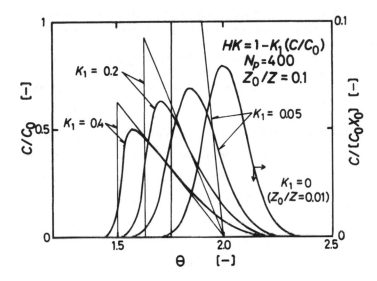

Fig. 2.4   Calculated elution curves by the equilibrium theory as a function of nonlinearity of the distribution coefficient. The concentration dependence of the distribution coefficient K is expressed by $HK = 1 - K_1(C/C_0)$, where K_1 is the parameter expressing nonlinearity. Thin curves are calculated by the equilibrium theory, Eq. (2.20). Thick curves are numberial solutions of the continuous-flow plate theory, Eq. (2.34), described in Sec. 2.1.4. The calculation scheme is almost the same in Appendix D. Z_0 is the length from the top of the column in which the solute is present before elution. The curve for $K_1 = 0$ (linear isotherm) is calculated with $Z_0/Z = 0.01$ and the concentration is expressed as $C/[C_0 X_0]$, whereas for the other curves Z_0/Z are taken to be 0.1 and the concentration is given as C/C_0.

2.4 are often observed not only in isocratic but also in gradient elution, as shown in Sec. 8.1.4, when the solute follows a nonlinear isotherm.

2.1.3   Mass Balance Model (Rate Theory)

During the elution process, a solute zone broadens because of several factors. In general, zone spreading can be considered to be caused by the following two mechanisms:

1. Zone spreading in the mobile phase (the interparticle space)
    is caused by eddy diffusion (velocity inequalities) and mo-
    lecular diffusion.
2. The diffusion inside a particle and the stagnant fluid film
    around the particle play a dominant role in zone spreading
    due to the finite transport rate between the mobile and the
    stationary phases.

In the theoretical treatment of the equilibrium theory shown
in the previous section, zone spreading effects are ignored. If
we take zone spreading effects into consideration, a more rigorous model, called a mass balance model or a rate theory, is derived. When axial dispersion can be represented by the effective longitudinal dispersion coefficient D_L mass balance in a
small element in a column gives (see Fig. 2.5).

$$\frac{\partial C}{\partial t} = D_L \frac{\partial^2 C}{\partial z^2} - u \frac{\partial C}{\partial z} - H \frac{\partial \overline{C}_{av}}{\partial t} \tag{2.21}$$

where $\overline{C}_{av} = (3/R_p^3) \int_0^{R_p} r^2 \overline{C}\, dr$ = average value of $\overline{C}$

$R_p = d_p/2$

R_p, d_p = particle radius and diameter, respectively

r = radial distance from the center of the particle (see Fig.
    2.6)

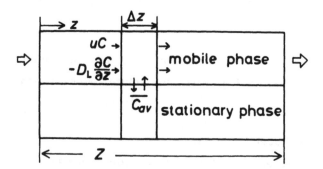

Fig. 2.5 Mass balance in a column.

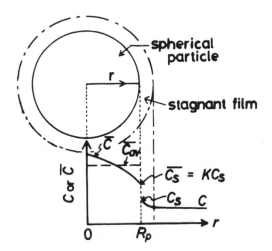

Fig. 2.6   Schematic representaiton of concentration profiles inside and outside a spherical particle.

Equation (2.21) is the basic partial differential equation for the
mobile phase.   The term on the left-hand side represents the
accumulation of a solute in the mobile phase; the terms on the
right-hand side describe the transport of the solute by longitudinal dispersion by convection and by the volume-averaged
accumulation of the solute in the gel phase.   The diffusion inside
a spherical porous particle, which is regarded as a single homogeneous medium, is given by the diffusion equation

$$\frac{\partial \overline{C}}{\partial t} = \overline{D}\left(\frac{\partial^2 \overline{C}}{\partial r^2} + \frac{2}{r}\frac{\partial \overline{C}}{\partial r}\right) \qquad (2.22)$$

where $\overline{D}$ = effective gel-phase diffusion coefficient.   (See Chap.
4 for the details of the diffusion model in the porous particle.)
Transport of the solute in the stagnant film around the particle
is described as

$$\frac{\partial \overline{C}_{av}}{\partial t} = k_f \frac{3}{R_p} (C - \overline{C}_s/K) = K_f(KC - \overline{C}_s) \qquad (2.23)$$

where $K_f = k_f(3/R_p)/K$

$\overline{C}_s = KC_s$

k_f = mass transfer coefficient

$\overline{C}_s$ = concentration at the interface between the particle and the outer solution (see also Fig. 2.6)

K_f = mass transport coefficient defined by Eq. (2.23).

Equation (2.23) can be used as the link between Eqs. (2.21) and (2.22). The initial and boundary conditions for the mobile phase are the same as those described in the equilibrium theory (see Sec. 2.1.2). Since Eq. (2.21) includes the second derivative of C with respect to z, an additional boundary condition is needed. The most commonly employed boundary condition is $C = 0$ at $z = \infty$. The initial condition for Eq. (2.22) is $C = 0$ at $t = 0$. One of the boundary conditions is $\partial \overline{C}/\partial r = 0$ at $r = 0$; another is given by Eq. (2.23) at $r = d_p/2 = R_p$.

The preceding set of equations is the most rigorous and complicated mass balance model, which was first employed by Deisler and Wilhelm (1953) for the study of diffusion processes in fixed-bed catalytic reactors. This model is referred to here as model I. Before describing the solution of this model, we give several models obtained by simplifying model I. This is because such simplified models are much easier to use and their analytical solutions are also simple. However, we should keep in mind that in employing such simplified models and/or their analytical solutions we must examine whether the assumptions involved in the simplification are valid in respective cases. Some discussion of this is presented in the next section.

Let us make the following assumptions.

Assumption 1   The diffusion inside the particle is so rapid that there is no concentration gradient inside the particle; that is, $\overline{C}_{av} = \overline{C}_s$.

Assumption 2   The stagnant film is negligible.   Namely, K_f is infinite.   Then, Eq. (2.23) becomes

$$\frac{\partial \overline{C}_{av} / \partial t}{K_f} = KC - \overline{C}_s$$

$$0 = KC - \overline{C}_s \tag{2.23'}$$

$$\overline{C}_s = KC$$

Assumption 3   The longitudinal dispersion is negligibly small; that is, $D_L = 0$.

If we employ all these assumptions, Eq. (2.21) reduces to Eq. (2.6), the basic equation of the equilibrium theory.   Therefore, we can regard the equilibrium theory as one of the limiting forms of the mass balance model.

The second simplest model, which we call model II, is obtained with assumptions 1 and 2.   The basic equation is given by the equation

$$\left[1 + H \frac{\partial \overline{C}}{\partial C} \right] \frac{\partial C}{\partial t} = D_L \frac{\partial^2 C}{\partial z^2} - u \frac{\partial C}{\partial z} \tag{2.24}$$

With constant K (linear chromatography), this equation is rewritten with dimensionless variables as

$$\frac{\partial C}{\partial \theta} = \frac{R}{(2N_p)} \frac{\partial^2 C}{\partial z*^2} - R \frac{\partial C}{\partial z*} \tag{2.25}$$

where $z* = z/Z$, $R = 1/(1 + HK)$ and $N_p = Z/[2(D_L/u)]$.   The meaning of the variable N_p is discussed in Sec. 2.1.5.   The analytical solution of Eq. (2.25) has been derived by several authors for both a pulse input (elution curve) and a step input (breakthrough curve) (Lapidus and Amundson, 1952; Kubin, 1965; Grubner and Underhill, 1972). The solution for a small amount of a solute (pulse input) is given by

$$C = C_0 X_0 \left(\frac{N_p}{2\pi R \theta^3} \right)^{1/2} \exp \left[-\frac{N_p R}{2\theta} \left(\theta - \frac{1}{R} \right)^2 \right] \qquad (2.26)$$

The solution of Eq. (2.25) becomes somewhat different from that of Eq. (2.26) when the following solution procedure is applied. Let define a new moving frame, $\xi = z^* - R\theta$. With this new coordinate system, Eq. (2.25) is transformed to the following simple equation, which has the same form as the one-dimensional diffusion equation (Fick's second law).

$$\frac{\partial C}{\partial \theta} = \frac{R}{(2N_p)} \frac{\partial^2 C}{\partial \xi^2} \qquad (2.27)$$

Once this equation is obtained, we can use several solutions compiled by Crank (1975). The solution of the pulse input by this procedure becomes

$$C = C_0 X_0 \left(\frac{N_p R}{2\pi \theta} \right)^{1/2} \exp \left[-\frac{N_p R}{2\theta} \left(\theta - \frac{1}{R} \right)^2 \right] \qquad (2.28)$$

Under the usual liquid chromatographic conditions, these two solutions are almost identical, as shown later in Fig. 2.10.

The exact analytical solution for the variable distribution coefficient has not yet been obtained. When the distribution coefficient is expressed by a linear function of concentration and its nonlinearity is small, approximate analytical solutions for a pulse input are obtained (Houghton, 1963; Haarhoff and Van Der Linde, 1966). Numerical solutions are also presented by several researchers (Dunckhorst and Houghton, 1966; Poppe and Kraak, 1983). When the distribution coefficient is a decreasing function of concentration, the elution curve given by the preceding solutions has a sharp front and a spreading boundary (tail); its limiting form is described by the curves shown in Fig. 2.4.

The third model, model III, which consists of the set of equations (2.21) and (2.23), is given with assumption 1. We can replace Eq. (2.23) with several other expressions with different physical meanings that have the same form as Eq. (2.23). Although such expressions may be contradictory to assumption 1, we include them in model III since the mathematical treatment is the same. We survey such expressions very briefly.

First-Order Reversible Surface Adsorption-Desorption

When the desorption and adsorption rate constants are given by K_d and K_a, respectively, the following equation is obtained:

$$\frac{\partial \overline{C}_{av}}{\partial t} = K_a C - K_d \overline{C}_{av} = K_d (KC - \overline{C}_{av}) \tag{2.29}$$

where $K = K_a/K_d$.

Linear Driving Force Approximation (Glueckauf-Coates Equation)

The equation in which the driving force for the internal mass transfer in the particle is assumed to be the difference between the concentration at the interface $\overline{C}_s = KC$ and the average concentration $\overline{C}_{av}$ is given by

$$\frac{\partial \overline{C}_{av}}{\partial t} = K_s(KC - \overline{C}_{av}) \tag{2.30}$$

where K_s = mass transfer coefficient, which depends on time and position in the elution process. The value of K_s is discussed in Sec. 2.1.5. This simplified treatment was first introduced by Glueckauf and Coates (1947). Detailed analysis of the linear driving force equation has been reported (Hills, 1986; Do and Rice, 1986).

Internal and External Mass Transfer

If we assume that both K_s and K_f contribute to the mass transfer rate, the following equation is obtained:

$$\frac{\partial \overline{C}_{av}}{\partial t} = K_t(KC - \overline{C}_{av}) \tag{2.31}$$

where K_t is the overall mass transfer coefficient and is given by $K_t = 1/(1/K_f + 1/K_s)$. It is quite interesting that Eqs. (2.23), (2.29), (2.30), and (2.31) can be rewritten in a generalized form as

$$\frac{\partial \overline{C}_{av}}{\partial t} = K_n(KC - \overline{C}_{av}) \tag{2.32}$$

where suffix n can be d, f, s, or t. Thus the set of equations composed of Eq. (2.21) and Eq. (2.32) comprise model III.

This set of equations was solved analytically in the case of the constant distribution coefficient by Lapidus and Amundson (1952) for both the pulse and step inputs. Their solution for the pulse input was later approximated to a gaussian solution by van Deemter et al. (1956). Houghton (1964) has shown that when the value of K_n is not very large (the condition needed to

achieve this requirement are discussed in Sec. 2.1.5), K_n can
be incorporated into the term N_p as

$$\frac{1}{N_p} = \frac{2(D_L/u)}{Z} + \frac{2R(1 - R)}{\tau K_n} \qquad (2.33)$$

Model III is thus reduced to model II. Therefore the solution,
Eq. (2.26) or Eq. (2.28), can be directly employed, and finally
the same solution as that of van Deemter et al. (1956) is obtained. Analytical solutions of model III for variable K have
not been obtained owing to the mathematical complexities.

We then return to model I, the most rigorous but complicated
model. Analytical solutions in the case of the constant distribution coefficient for the step and pulse inputs have been derived
by Rasmuson and Neretnieks (1980) and Wakao and Tanaka
(1973), respectively. However, since these analytical solutions
are extremely complex, it is difficult to extract explicit relations
between a particular variable and the elution behavior. A better
way to obtain the relations between the operational variables
and the nature of the elution curve is to use the moment
method, as described in Sec. 2.1.1. Kubin (1965) and Kucera
(1965) derived moments of the elution curve from the solution
in the Laplace domain of model I. (The moments they derived
are given in Sec. 2.1.5.) Since then, model I has sometimes
been called the Kubin-Kucera model. Once the moments are obtained, they can be employed for the prediction of the elution
behavior when the values of such parameters as $\overline{D}$, D_L, K_f, and
K are known and for the determination of the parameters from
the elution curves. The latter application is presented in Chaps.
3 through 5. Calculation of the analytical solutions requires
much computing time, as pointed out by Raghavan and Ruthven

(1983). They compared the computing time of the analytical solution for the breakthrough curve with that of the numerical solution by the method of orthogonal collocation and found that the latter method requires less computing time. On the other hand, we employed the numerical inversion of the Laplace transform (Crump, 1976; Dubner and Abate, 1968) to calculate the elution curve by model I (Yamamoto et al., 1979; Nakanishi et al., 1979). This method is very simple and rapid.

Application of model I to the variable distribution coefficient case is so complex that it is not easy to obtain numerical solutions. Rather, it is advisable to employ model II for the analysis of liquid chromatographic separation processes on the basis of such knowledge of the range of the applicability of model II as that given in Sec. 2.1.6.

Although many approximate or numerical solutions are obtained from the mass balance model, which is set up with the assumptions that the axial dispersion is negligible (assumption 3) and that the linear driving force equation can be employed for mass transport inside the stationary phase, they are not treated here. Such solutions are usually obtained for the breakthrough curve (step input), not for the elution curve (pulse input). Readers who are interested in this field can find a number of important papers on the fixed bed (for example, Vermeulen et al., 1984) and on affinity chromatography (Yang and Tsao, 1982; Arnold et al., 1985b; 1985c; Chase, 1984a; 1984b; Katoh and Sata, 1980a; 1980b; Sada et al., 1982; 1984; 1985).

2.1.4 Plate Theory

The similarity of the process occurring in chromatography to that in distillation columns was first observed by Martin and Synge

(1941). They proposed a mathematical model called the plate
theory in which a column is regarded as a series of plates or
well-mixed tanks. In their original treatment, the first plate
is assumed to be occupied with the sample before elution. The
solution of this model was given by Poisson distribution, which
can be approximated by the gaussian curve for larger theoretical
plates. Mayer and Tompkins (1947) also reported a similar approach except that the flow is discontinuous (a flow increment is
assumed equal to the volume of the mobile phase of a theoretical
plate). Glueckauf (1955a,b) pointed out that the Mayer and
Tompkins discontinuous-flow model would give large errors even
if a column had 1000 theoretical plates. He then derived a partial differential equation on the basis of the continuous-flow
model, assuming a large number of theoretical plates, and obtained a solution for the equation. Said (1956) also set up a
set of ordinary differential equations based on the continuous-
flow plate theory and solved the equation for various initial and
boundary conditions. Reilley et al. (1962) obtained solutions of
the Mayer and Tompkins discontinuous-flow plate theory for
various initial and boundary conditions. We first describe the
continuous-flow plate theory by using the same treatment as
Said (1956) and van Deemter et al. (1956).

A basic equation for the nth plate is given by the following
differential equation. (Since our model for ion-exchange chromatography of proteins is based on this model, the details of
the model are described in Sec. 2.2.2. Therefore, only the
basic equation is given here.)

$$\frac{dC_{(n)}}{d\theta} + \frac{Hd\bar{C}_{(n)}}{d\theta} = N_p[C_{(n-1)} - C_{(n)}] \qquad (2.34)$$

where $C_{(n)}$ and $\overline{C}_{(n)}$ = solute concentration in the mobile and stationary phases of the nth plate, respectively, and N_p = total number of theoretical plates of the column. When the distribution coefficient is constant, Eq. (2.34) is rearranged as

$$\frac{dC_{(n)}}{d\theta} = RN_p[C_{(n-1)} - C_{(n)}] \qquad (2.34')$$

The assumptions common to various plate theories are as follows:

1. The column consists of a certain number of equivalent theoretical plates, for which the ratio of the volume of the stationary phase to that of the mobile phase is the same.
2. The equilibrium of solutes between the two phases is instantaneously attained.

The solution for the nth plate when the first plate is occupied by the solute before elution (the Martin and Synge case) with constant K is given by the Poisson distribution function

$$C_{(n)} = C_0(RN_p\theta)^{n-1}\frac{\exp(-RN_p\theta)}{(n-1)!} \qquad (2.35)$$

This equation is reduced to the following gaussian function when the number of theoretical plates N_p is sufficiently large (van Deemter et al., 1956):

$$C = C_0 X_0 \left(\frac{N_p R}{2\pi\theta}\right)^{1/2} \exp\left[-\frac{N_p R}{2\theta}\left(\theta - \frac{1}{R}\right)^2\right] \qquad (2.36)$$

Note that this equation has exactly the same form as that of Eq. (2.28).

The discontinuous-flow plate theory is based on the discontinuous increment of liquid flow equal to the volume of the mobile phase of one plate in addition to assumptions 1 and 2. The solu-

tion at the outlet of the column is represented by the binomial
distribution function

$$C = \frac{C_0 X_0 (N_p \theta)!}{(N_p \theta - N_p)! N_p!} R^{N_p} (1 - R)^{N_p \theta - N_p} \tag{2.37}$$

Correspondence of the plate theory with the mass balance model
has been demonstrated by Giddings (1965) and Glueckauf (1955a,
b). We show the correspondence using an approach similar to
theirs. When the N_p value is large and the zone is developed
fully, the first and second terms on the right-hand side of
Eq. (2.25) are replaced by their central difference approximation
with its distance of the grid Δz equal to one theoretical plate
height Z/N_p; that is, $\Delta z^* = 1/N_p$. Equation (2.25) then reduces
to Eq. (2.34'), as follows:

$$\frac{\partial C_{(n)}}{\partial \theta} = \frac{R}{2N_p} \frac{[C_{(n+1)} - C_{(n)}] - [C_{(n)} - C_{(n-1)}]}{\Delta z^{*2}}$$

$$- \frac{R[C_{(n+1)} - C_{(n-1)}]}{2\Delta z^*}$$

$$= \frac{RN_p}{2} [C_{(n+1)} - 2C_{(n)} + C_{(n-1)} - C_{(n+1)} + C_{(n-1)}]$$

$$= RN_p [C_{(n-1)} - C_{(n)}]$$

The plate theory concept is not rigorous for describing the
actual chromatographic process. Since N_p is not directly related
to the operating variables, it is impossible to predict how a
particular operating variable affects the elution behavior. In
spite of these disadvantages, the concept of a height equivalent
to a theoretical plate (HETP) derived from plate theory has gained
its utility and has been employed as a measure of column efficiency.

Moreover, as shown in the following section, a number of
equations have been derived that relate HETP to the column and
operating variables. Plate theory has been recently employed
again since it can be easily applied to nonlinear chromatographic
processes that cannot be easily treated by the mass balance
model because of mathematical difficulties (Wankat, 1974a; Nelson
et al., 1978; Yamamoto et al., 1983a, b). The elution curves
for nonlinear isotherms were calculated by continuous-flow plate
theory. The results are shown in Fig. 2.4. It is seen that the
calculated curves are broadened compared with those obtained
by the equilibrium theory. Comparison of plate theory with the
mass balance model is made in the next section in terms of the
moment equations.

2.1.5 HETP and Plate Number

In continuous-flow plate theory, the elution curve can be expressed by the gaussian distribution function when the number
of theoretical plates N_p is large, as stated in the previous section. The following relation is then derived:

$$N_p = \frac{t_R{}^2}{\sigma^2}$$

$$= \frac{(1 + HK)^2}{\sigma_\theta{}^2} \tag{2.38}$$

where σ^2 and $\sigma_\theta{}^2$ = variance of the elution curve with respect
to time and to dimensionless time θ, respectively, and t_R = peak
position of the elution curve. A height equivalent to a theoretical plate is given here

$$HETP = \frac{Z}{N_p} = Z \frac{\sigma^2}{t_R^2} \tag{2.39}$$

According to these equations, we can determine the value of HETP or N_p from experimental elution curves. However, these equations are not explicit functions of operating and column variables. The first attempt to relate HETP to such variables was made by van Deemter et al. (1956). As mentioned in Sec. 2.1.3, they approximated the solution of the mass balance model (model III) presented by Lapidus and Amundson (1952) to the gaussian distribution function and derived an HETP equation. Their HETP equation is given with our notation as

$$HETP = 2 \frac{D_L}{u} + 2u \frac{R(1 - R)}{K_s} \tag{2.40}$$

Van Deemter et al. (1956) considered that D_L can be represented by the sum of the effect of the eddy diffusivity and the molecular diffusivity.

$$D_L = \lambda u d_p + \gamma_m D_m \tag{2.41}$$

where λ = packing characterization factor

d_p = particle diameter

γ_m = obstructive (tortuousity) factor in the intraparticle space

D_m = molecular diffusivity

The value of K_s was also derived from the analytical solution of the diffusion equation in a sphere with a constant surface concentration (Crank, 1975).

$$K_s = \frac{4\pi^2 \bar{D}}{d_p^2} \tag{2.42}$$

Insertion of these two equations into Eq. (2.40) yields a well-known HETP equation, the van Deemter equation.

$$HETP = A + \frac{B}{u} + Cu \tag{2.43}$$

where $A = 2\lambda d_p$

$B = 2\gamma_m D_m$

$C = (1/2\pi^2)R(1 - R)d_p^2/\overline{D}$

Next we describe an HETP equation derived from the moment method. As stated in Sec. 2.1.1, the moments of the elution curve can be easily related to the parameters appearing in the model with the solution of the partial or ordinary differential equation in the Laplace domain $\tilde{C}(p)$ of the model (Kubin, 1965; Kucera, 1965; Grubner and Underhill, 1972; Grushka, 1972; 1975). The moments of the elution curve for various models are summarized in Table 2.1, where the values of N_p defined by the following equation are also given.

$$N_p = \frac{(\mu_1')^2}{\mu_2} \tag{2.44}$$

HETP is then given by

$$HETP = \frac{Z}{N_p} = \frac{Z\mu_2}{(\mu_1')^2} \tag{2.45}$$

An HETP equation for the mass balance model of model I (the Kubin-Kucera model), in which the term including K_f is omitted, is expressed as

$$HETP = \frac{2D_L}{u} + \frac{HKd_p^2 u}{30\overline{D}(1 + HK)^2}$$

$$= \frac{2D_L}{u} + \frac{(1/30)R(1 - R)d_p^2 u}{\overline{D}} \qquad (2.46)$$

This equation is rewritten in terms of the dimensionless variables.

$$h = \frac{2}{Pe} + \frac{(1/30)R(1 - R)\nu}{\gamma_{sm}} \qquad (2.47)$$

where h = HETP/d_p = reduced HETP

$Pe = d_p u/D_L$

$\gamma_{sm} = \overline{D}/D_m$

$\nu = d_p u/D_m$ reduced velocity

It is interesting to note that Eq. (2.46) has the same form as the van Deemter equation. The only difference is the numerical value in ther C term. This is because van Deemter et al. (1956) employed the solution for constant surface concentration for the determination of K_s. If the solution for variable surface concentration is used, the more rigorous value of K_s is obtained. This was done by several authors (Glueckauf, 1955; Bogue, 1960; Huang et al., 1983). The values in these reports are the same and are given by Eq. (2.48), although the methods for the derivation are different. The same result as Eq. (2.48) was obtained by using the properties of the Laplace transform (see Appendix A).

$$K_s = \frac{60\overline{D}}{d_p^2} \qquad (2.48)$$

Introduction of this equation into Eq. (2.40) yields an equation that is completely identical to Eq. (2.46).

Table 2.1 Equations for the Number of Theoretical Plates N_p, Skewness, and Kurtosis from Various Chromatographic Models[a]

Model	N_p[b]	Skewness[c]	Kurtosis[c]
Continuous-flow plate theory	N_p	$2/N_p^{1/2}$	$6/N_p$
MBM model I[d]	$\dfrac{Z}{2\dfrac{D_L}{u} + \dfrac{HKd_p^2 u}{30(1+HK)^2\bar{D}}}$	at $D_L \to 0$ $\dfrac{15(1+HK)}{7HK\,(N_p^{1/2})}$	at $D_L \to 0$ $\dfrac{45(1+HK)^2}{7(HK)^2 N_p}$
MBM model II	$\dfrac{Z}{2D_L/u}$	$\dfrac{3}{N_p^{1/2}}$	$\dfrac{15}{N_p}$

MBM model III	$\dfrac{K_n(1+HK)^2\tau_e}{2HK}$	$\dfrac{3(1+HK)}{2HK(N_p^{1/2})}$	$\dfrac{3(1+HK)^2}{(HK)^2 N_p}$

[a] The first statistical normalized moment (the mean retention time) for all the models is given by $\tau(1+HK)$; the second central moment (the variance) is $\tau^2(1+HK)^2/N_p$. The variance of the discontinuous-flow plate theory is $\tau^2(1+HK)HK/N_p$.

[b] N_p is obtained according to Eq. (2.44).

[c] Skewness and kurtosis approach zero when the elution curve is close to the gaussian error curve.

[d] MBM is the mass balance model. In MBM model I, the term that considers the mass transfer through a stagnant film around the particle, the D term, is omitted in the N_p equation. In addition, since the complete expressions for the skewness and the kurtosis of MBM model I are extremely complex, only the limiting solutions at $D_L \to 0$ are given. When $\bar{D}$ is infinite, the skewness and kurtosis of model I reduce to those of model II. Complete expressions for the N_p, skewness, and kurtosis of model I are found in the literature (Kubin, 1965; Kucera, 1965).

[e] K_n (n = d, f, s, t) is a general form of the mass transfer coefficient. When $K_f = (3/R_p)k_f/K$ is inserted into K_n, the resulting N_p becomes the D term in model I. It is also noteworthy that the insertion of Eq. (2.48) into K_n gives the N_p term when only the stationary-phase diffusion is a rate-controlling step.

It is stressed here that although various mass balance approaches are available for the description of chromatographic
processes, the resulting HETP equation leads to Eq. (2.46) if
each factor affecting zone spreading occurs independently and is
additive (Arnold et al., 1985a, c). In addition, many HETP
equations in the literature (for example, Engelhardt and Ahr,
1983; Horvath and Lin, 1976, 1978; Walters, 1982; Guiochon and
Martin, 1985; Ghrist et al., 1987) are considered extended or
modified forms of Eq. (2.46), as demonstrated by Katz et al.
(1983). For example, the Giddings HETP equation, derived from
his coupling theory, converges to Eq. (2.46) at high linear
velocities (Giddings, 1975). Katz et al. (1983) tested various
HETP equations, including the van Deemter equation, against a
number of experimental data on HETP as a function of flow rate
for silica gel columns and found that the van Deemter equation
most accurately represents the relationship between HETP and u.
Knox and Scott (1983) examined the B and C terms in the van
Deemter equation in detail and reported that over a limited velocity range the van Deemter equation is adequate. However,
they concluded that the most appropriate equation is the Kennedy
and Knox equation (Kennedy and Knox, 1972), in which the A
term in the van Deemter equation is not constant but varies
with the one-third power of the linear velocity. At low ν values
($\nu < 1$), that is, when the particle diameter d_p and/or u are
very small and/or the molecular diffusion coefficient D_m is large,
the coupling phenomena as described by Giddings (1965) may
occur. In such a case the A term is not constant but varies
with ν. However, as described later for the liquid chromatography of proteins, the value of ν ranges between 20 and 200
owing to the low D_m values. In this range of ν, the coupling

effect is not observed (Arnold et al., 1985a), the A term can be taken to be constant, and the B term is negligible, as shown in Fig. 2.12 and described in Chap. 5. HETP can then be expressed simply by HETP = A + Cu or h = A + Cν. In other words, HETP is a linear function of u or ν. It is considered that this equation is most suitable for expressing HETP versus u relations for the liquid chromatography of proteins.

However, in certain cases, empirical HETP equations may be employed for relating the observed HETP to u. For example, Snyder's equation (Snyder and Saunders, 1969), HETP = Du^n or h = $Dν^n$ (D and n are constants for a given column), is frequently employed to fit experimental data. An advantage of the use of this equation for fitting the experimental data is its simplicity, although the equation itself is empirical. For instance, if this equation is inserted into the R_s equation, we can see very easily how R_s depends on d_p and/or u (for example, Snyder and Kirkland, 1974).

It is assumed very frequently in liquid chromatography, especially that of proteins, that the term describing the contribution of the interfacial mass transfer coefficient k_f to the total HETP (this term will now be called the D term) can be omitted. The D term in the HETP equation from the mass balance model I is given as $(1/30)R(1 - R)(10K/Sh)ν$ in the reduced form, where Sh = $k_f d_p/D_m$ is the Sherwood number. Therefore, we can see from the second term in Eq. (2.47) that comparison of the 10K/Sh value with the $1/\gamma_{sm}$ value gives the relative contribution of the D term to the C term. Let us employ the following values: $\gamma_{sm} = \overline{D}/D_m = 0.1$ (see Chap. 4), K = 0.5, ε = 0.38, and ν = 20 (a very low value in the actual liquid chromatography of proteins, as shown in Sec. 2.1.6). As for the

correlation of Sh, the following equation has been proposed for
the packed bed by Wilson and Geankoplis (1966).

$$Sh = 1.09\epsilon^{-2/3}\nu^{1/3} \tag{2.49}$$

Insertion of these values into this equation yields Sh = 5.64.
We see that the D term is about 0.089 of the C term. With an
increase in ν, the contribution of the D term to the C term will
diminish. Therefore, we may neglect the D term (k_f term)
under the usual chromatographic conditions for proteins.

It is difficult to derive an HETP or N_p equation for nonlinear
chromatography. Buys and DeClerk (1972a, b) and DeClerk and
Buys (1971) obtained approximate expressions for moments from
the same mass balance equation and an expression for the nonlinearity of the isotherm as that employed by Houghton (1963). Poppe
and Kraak (1983) also treated the same model and compared the
first and the second moments of numerical solutions with those
from the approximate solution of Haarhoff and van der Linden
(1966).

Several expressions have been proposed for the column
efficiency other than HETP (or N_p). The number of effective
plates N_{eff} given by the following equation is frequently used
since it simplifies the R_s equation (2.54):

$$N_{eff} = \frac{[t_R - (z/u)]^2}{\sigma^2}$$

$$= \left(\frac{HK}{1 + HK}\right)^2 N_p \tag{2.50}$$

Snyder and Kirkland (1974) related N_{eff} to two important factors: the separation time and the pressure drop. In linear

isocratic chromatography, the separation time can be calculated
by the first and second moment equations. The pressure drop
per unit length with a rigid spherical particle gel column is
estimated with the aid of the Kozeny-Carman equation (see Chap.
7). However, the experimental conditions for obtaining a desired N_p must be determined by trial-and-error calculation,
since the increase in N_p is always accompanied by an increase
in either the separation time or the pressure drop.

2.1.6 Experimental Verification of HETP

In the previous section, we described that the HETP equation,
Eq. (2.46), is the most rigorous for liquid chromatography,
especially for proteins. However, it should be kept in mind
that even if the values of HETP or N_p are the same, the shape
of the elution curve is not always identical. It is known that
the elution curve becomes skewed to a smaller elution time and
has a long tail when the particle diameter is large, the flow
rate is high, and the solute has a low molecular diffusivity
(Yamamoto et al., 1978, 1979). Typical examples of such elution curves are shown in Fig. 2.7. With an increase in the
flow rate, the elution curve shifts its peak position to the lower
elution time and the length of the tail increases. This is not
only unfavorable for good separation but also confusing in in-
terpreting the experimental results. The degree of the shift
of peak position, which corresponds to the magnitude of the
peak asymmetry, was correlated with three dimensionless variables. For this purpose, the elution curves for the mass
balance model (model I, the Kubin-Kucera model) were calculated
numerically by the numerical Laplace transform method (Dubner
and Abate, 1968; Crump, 1976). The results are shown in Fig.

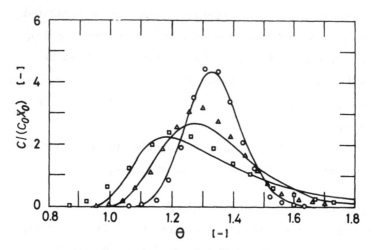

Fig. 2.7 Observed and calculated elution curves for myoglobin with gel filtration column as a function of flow rate. Column, Sephadex G-50 (d_p = 202 μm, d_c = 1.5 cm, Z = 15 cm). The solid curves are the calculated results from model I with $\overline{D}$ = 9.5 × 10^{-8} cm^2/s, D_L/u = 0.0202 cm, and HK = 0.34. F(=mL/min); □ = 1.0, △ = 0.44, ○ = 0.1. (From Yamamoto et al., 1979.)

2.8 (Yamamoto et al., 1978; Nakanishi et al., 1983). The decrease of K_{peak}/K, which implies the increase of the magnitude of peak asymmetry, is caused by the increase of the dimensionless variable $\phi = R_p(\tau/\overline{D})^{1/2}$. Furthermore, K_{peak}/K is sensitive when HK is small and/or the dimensionless variable $\delta = Z/(D_L/u)$ is large. From these findings, we expect that the following factors cause the peak asymmetry: the increase in d_p, the decrease in residence time, and the decrease in molecular diffusivity. This is the origin of the tailing due to the low mass transport rate between the mobile and the stationary phases, which must be distinguished from that due to the nonlinear isotherm, as shown in Sec. 2.1.2. Simpler variables should be used for the prediction of the peak asymmetry. Another correlation in terms of N_p was

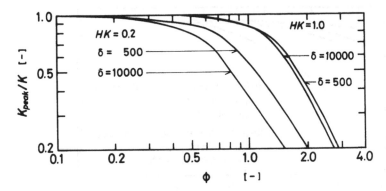

Fig. 2.8   Relation between K_{peak}/K and ϕ.   (From Yamamoto et al., 1979.)

therefore attempted (Fig. 2.9). From Fig. 2.9, we can easily predict the conditions in which the peak asymmetry is small. In general, when N_p is above 100, the elution curve is almost symmetrical and the curves calculated by different models are hardly distinguishable, as shown in Fig. 2.10. Grubner and Underhill (1972) reached a similar conclusion.

Several experimental HETP values for gel filtration and ion-exchange chromatography under conditions such that the peak is symmetrical (described earlier) are shown in Fig. 2.11. IEC is operated isocratically at high ionic strengths at which the electrostatic interaction between the ion-exchange group and proteins is negligible. Consequently, these ion-exchange gels act as gel filtration chromatographic media. Each set of HETP versus u data is expressed by a linear function of linear mobile-phase velocity u with different slopes and intercepts. This implies the applicability of Eq. (2.46). These experimental re-sults together with the additional data are replotted in Fig. 2.12 in the reduced forms, where h = $HETP/d_p$ and $\nu = ud_p/D_m$.

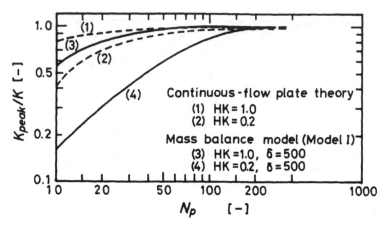

Fig. 2.9   Relation between K_{peak}/K and N_p.   (From Nakanishi et al., 1983.)

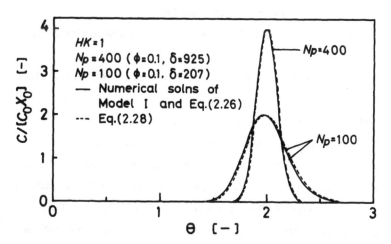

Fig. 2.10   Elution curves calculated by different models.

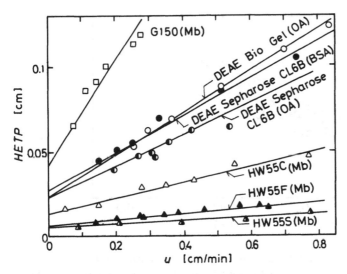

Fig. 2.11 HETP versus linear mobile-phase velocity u for various gel filtration columns and for ion-exchange gel columns at
high ionic strength where the ion-exchange gels act as gel filtration chromatographic media: Mb, OA, and BSA, myoglobin,
ovalbumin, and bovine serum albumin, respectively; G150,
Sephadex G-150 column (d_p = 202 μm, d_c = 1.5 cm, Z = 13 cm)
(from Nakanishi et al., 1978); HW55C, Toyopeark (TSK-GEL)
HW55C column (d_p = 75 μm, d_c = 1.6 cm, Z = 30 cm); HW55F,
Toyopearl HW55F column (d_p = 44 μm, d_c = 1.6 cm, Z = 30 cm);
HW55S, Toyopearl HW55SF column (d_p = 35 μm, d_c = 1.6 cm,
Z = 30 cm). (Data from Yamamoto et al., 1987a.) These gels
are gel filtration chromatographic media. DEAE-Sepharose GL6B
(d_p = 110 μm) and DEAE-Bio-Gel A (d_p = 123 μm) columns (d_c
= 1.5 cm, Z = 30 cm) were operated isocratically at ionic strength
= 0.5 M where proteins are not adsorbed onto the ion-exchange
gels and the gels act as gel filtration chromatographic media
(from Yamamoto et al., unpublished data). The temperature was
20 or 25°C.

The dimensionless plots h versus ν are very useful for examining the influence of various factors, such as d_p, u, and
the molecular diffusivity D_m, as already suggested by several
researchers (Giddings, 1965; Grushka et al., 1975), since the
HETP values over a wide range of experimental conditions can

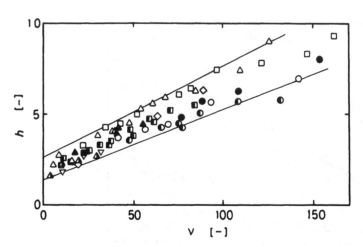

Fig. 2.12 Relation between h (= HETP/d_p) and ν (= ud_p/D_m).
All the experimental results shown in Fig. 2.11 are used. The
symbols are the same as in Fig. 2.11. The additional data are
for myoglobin on Toyopearl 55F columns (d_p = 44 μm, d_c = 1.6
cm, Z = 30 cm) at different temperatures ($\diamondsuit$ = 10°C, ∇ = 40°C)
and myoglobin ($\square$) on a TSK G3000SW high-performance gel-
filtration chromatography column (d_p = 11 μm, d_c = 0.75 cm,
Z = 30 cm). (These data are taken from Yamamoto et al.,
1987a.)

be compared in the same diagram. All the HETP values fell in

a narrow range encompassed by the two lines in Fig. 2.12, im-

plying that: (1) h versus ν relation is independent of d_p; (2)

the A term = ($2D_L/u$) is hardly dependent on d_p, temperature,

the type of proteins, or the type of gels and is approximated to

0.8–1.5d_p (see Chap. 5); and (3) the temperature dependence

of $\overline{D}_s$ is similar to that of D_m (see Appendix F for the method

of determining the temperature dependence of D_m). It should

be noted that there will be a minimum owing to the B/u term

(or D_m term) of Eq. (2.43) in the h versus ν relation when the

ν value is less than 5 (see also Fig. 5.1). Such conditions are

not usually encountered because of the low D_m values of proteins (the D_m of proteins is less than 1×10^{-6} cm^2/s at 20°C, as shown in Appendix F).

Since h is a unique function of ν for a given protein and gel type, the reduction of d_p drastically increases the separation efficiency. For example, when d_p is reduced from 100 to 20 µm, a fivefold increase of u gives the same ν value. The $N_p = Z/$ (HETP) = $Z/(hd_p)$ then increases by a factor of 5; that is, a fivefold decrease in d_p gives a fivefold increase in the separation efficiency in one-fifth the separation time. The column length can also be decreased to one-fifth of that with 100 µm particles to obtain the same N_p value. This is the reason that high-performance liquid chromatography (HPLC) packed with small particles gives a superior separation efficiency and a high speed of separation even when the column length is low (see Chap. 7 and Sec. 8.3.4). Guiochon and Martin (1985) have examined the optimum particle diameter for high-performance gel filtration chromatography of proteins using theoretical calculations.

It should be mentioned that the velocity range in the chromatography of proteins is limited to a low level compared with that of small molecules because of the low mechanical stability of the packing materials and the low stationary-phase diffusion coefficient. As described in Chap. 7, since most gel matrices employed for the protein separation are soft and compressible, there is a maximum flow rate for a given gel column. The slow diffusion velocity of proteins in the gel (the stationary phases), discussed in Chap. 4, is also a factor limiting the mobile-phase velocity, since the flow rate must be reduced in order to avoid the peak asymmetry due to such low gel-phase diffusivities as shown earlier. Therefore, even in high-performance gel filtra-

tion chromatography of proteins, in which the packing materials
(gels) are rigid, it is not usual that the dimensionless flow rate
ν exceeds 200–300. On the other hand, the HPLC of small molecules is operated between ν = 100 and 1000; sometimes ν is
beyond 1000. The narrow range of ν values also permits us to
assume that the D_L/u (the A term) for protein chromatography
is constant (see Chap. 5).

2.1.7 Resolution of Two Components

In the previous section, the zone spreading mechanism in liquid
chromatography was described in terms of HETP. We are now
able to predict the elution behavior of a given substance when
HETP is known as a function of u. However, for the prediction
of the separation efficiency of two components characterized by
HETP, we must consider the distance of the peaks of two adjacent elution curves in addition to their widths. As a measure
of the degree of separation efficiency of the two adjacent peaks,
the resolution R_s is frequently employed, defined by

$$R_s = \frac{t_{R,2} - t_{R,1}}{(W_1 + W_2)/2} \tag{2.51}$$

where t_R = peak position and W = width of the elution curve at
the baseline. The subscripts 1 and 2 refer to components
(peak) 1 and 2, respectively. When the elution curve can be
approximated to the gaussian curve, this equation is rewritten
in a dimensionless form as (see also Fig. 2.18).

$$R_s = \frac{\theta_{R,2} - \theta_{R,1}}{2(\sigma_{\theta,1} + \sigma_{\theta,2})} \tag{2.52}$$

where $\theta_R = t_R/(Z/u)$. It follows from Eq. (2.51) or (2.52) that
R_s increases with an increase in the distance between the two

peak positions and with an decrease in the peak width. In the case of linear isocratic chromatography, the peak position t_R can be expressed by $t_R = (Z/u)(1 + HK)$ under the usual chromatographic conditions and σ_θ^2 is given by $(1 + HK)^2/N_p$, as mentioned previously. We further assume that the peak width of substance 2 is almost equal to that of substance 1; that is, $\sigma_{\theta,2} = \sigma_{\theta,1}$. Equation (2.52) then becomes

$$R_s = \frac{H(K_2 - K_1)}{4[(1 + HK_1)^2/N_{p,1} + X_0^2/12]^{1/2}} \qquad (2.53)$$

The term $X_0^2/12$ appearing in the denominator is the zone spreading due to the initial sample volume (Kubin, 1965). If the sample volume is so small that the term $X_0^2/12$ becomes negligible compared with the term $(1 + HK_1)^2/N_{p,1}$, the equation becomes Eq. (2.54), which is commonly used for estimation of the separation efficiency (for example, Snyder and Saunders, 1969; Snyder and Kirkland, 1974; Saunders, 1975):

$$R_s = \frac{HK_1(K_2/K_1 - 1)}{4(1 + HK_1)/N_{p,1}^{1/2}} \qquad (2.54)$$

In order to relate R_s to operating variables, it is necessary to replace N_p by an appropriate equation that precisely describes N_p or HETP. As already stated, the HETP equation, Eq. (2.46), is considered most suitable. Insertion of Eq. (2.46) into Eq. (2.53) then yields the equation

$$R_s^{-2} = \frac{16[(1 + HK_1)^2 A + (1 + HK_1)^2 Cu + X_0^2 Z/12]}{[H(K_2 - K_1)]^2 Z} \qquad (2.55)$$

where $A = 2(D_L/u)$ and $C = (1/30)R(1 - R)d_p^2/\overline{D}$.

From this equation we can extract the following relations.

1. R_s increases with the square root of the column length.
    This is a well-known relation that holds irrespective of the
    type of HETP equation provided that the packing of longer
    columns is similar to that of smaller columns (Kato et al.,
    1981b).
2. As long as Eq. (2.46) is valid, there is a linear relation
    between R_s^{-2} and u.
3. Similarly, at a certain flow rate, R_s^{-2} is a linear function of
    X_0^2 (the square of the sample volume).

These relations indicate that in the isocratic elution chromatog-

raphy a long and narrow column gives a higher resolution than a

short and wide column when the volume of the two columns is the

same. A resolution equation for linear gradient elution is treated

in Sec. 2.2.4.

2.2 THEORY OF ION-EXCHANGE
CHROMATOGRAPHY OF PROTEINS

In the previous section, general theories of chromatography have

been surveyed. However, those theories are not directly ap-

plicable to the IEC of proteins, especially to gradient elution or

stepwise elution, which are the two most common elution tech-

niques because of their complicated elution mechanism. If we

treat the IEC of proteins as rigorously as possible according to

the description in Sec. 1.3, the elution of a protein and of a

salt must be considered coupled, not independent processes. In

addition, a pH change in the ion exchanger should also be taken

into account. However, this approach is extremely complicated

and difficult. Therefore, as mentioned in Sec. 1.4, we treat

the IEC of proteins in this book as simple adsorption chromatog-

raphy in which the elution of the salt and of the protein occur

independently, although the adsorption and diffusivity of proteins

are usually dependent on the salt concentration of the mobile
phase, as described in Chaps. 4 and 5. This treatment seems
to be reasonable since the diffusivity of salts is about 50 times
higher than that of proteins and this markedly reduces mathematical complexity. Basically, in this treatment we can obtain
the elution profile of a protein and of a salt by simultaneously
solving a mass balance equation for the protein and for the
salt. However, it is not an easy task even with this simplified
treatment. For example, when the mass balance model I described in Sec. 2.1.3 is applied to obtain the mass balance
equation, the resultant set of partial differential equations is
too complex to solve either analytically or numerically. This
complexity arises from the nonlinearity of the equations.

In this section, basic concepts needed for modeling the IEC
of proteins are first described in Sec. 2.2.1. Then, a model
proposed by the authors is presented (Yamamoto et al., 1983a).
This model is based on the continuous-flow plate theory (see
Sec. 2.1.4) and includes the protein concentration and ionic
strength dependencies of the distribution coefficient of protein
between the mobile and the stationary phases and also includes
zone spreading effects. Basic equations for both a linear
gradient elution and a stepwise elution derived from the model
and assumptions employed in its derivation are presented in
Sec. 2.2.2.

The main advantage of the model is its simplicity, since it
requires only two parameters: the distribution coefficient dependent on ionic strength and concentration and the number of
plates. The distribution coefficient can be determined by batch
experiments, shown in Sec. 3.2. As mentioned in Sec. 2.1.4,
the number of plates is such an important parameter that the

effects of all the factors that act on zone spreading are reflected
in it. However, since conventional methods for the determination of the number of plates, shown in Sec. 2.1.5, are not directly applicable, a method for determination is also presented
in Sec. 2.2.2. A calculation method for the elution profiles of
protein and salt by the model is also included in Sec. 2.2.2.

It is known that the peak position and width in a linear
gradient elution depend on both the slope of the gradient and
the column length (for example, Kato et al., 1982a). Although
these dependencies are predictable by the model in Sec. 2.2.2,
any simpler method would be helpful for the prediction and
optimization of the separation and for scaling up. A graphic
method is presented for prediction of the peak position on the
basis of the model proposed independently by Drake (1955) and
Freiling (1955). For prediction of the peak width, an asymptotic
solution is derived from a quasi-steady-state model. The
graphic method and asymptotic equation are described in Sec.
2.2.3.

R_S, which is defined by Eq. (2.51), is frequently employed
as a measure of the degree of the separation of two peaks. An
R_S equation for a linear gradient elution is also derived from a
quasi-steady-state model described in Sec. 2.2.3 and presented
in Sec. 2.2.4.

The advantages and disadvantages of the model and methods
presented are discussed in comparison with other models in
Sec. 2.2.5.

2.2.1 Basic Concepts

At equilibrium, the protein concentration in the stationary phase
$\overline{C}$ is related to that in the mobile phase C as a function of the

protein concentration and of ionic strength in the mobile phase, C and I:

$$\overline{C} = K[C, I]C \qquad (2.56)$$

where K is the protein concentration- and ionic strength-dependent distribution coefficient between $\overline{C}$ and C, which usually decreases with an increase in C and/or I (see Chap. 3).

In equilibrium theory, in which zone spreading effects are ignored, the moving rate of the protein zone in the column is expressed by the following equation (see also Sec. 2.1.2) (Drake, 1955; Freiling, 1955; DeVault, 1943; Weiss, 1943):

$$\frac{dZ_p}{dt} = \frac{u}{1 + HK[C, I]} \qquad (2.57)$$

where z_p is the peak position of the protein zone. Here, H is $(\overline{V}_t - \overline{V}_0)\overline{V}_0$, where V_t and V_0 are the total and void volumes of the column, respectively.

In linear gradient elution, the value of K is large at the beginning of the elution owing to low ionic strengths (see Chap. 3) and therefore dZ_p/dt is low. Since the linear increase in the ionic strength is continuously applied to the column, the value of K decreases. Therefore, the protein zone moves slowly at first and accelerates gradually with time. However, dz_p/dt approaches its maximum value a short distance from the top of the column owing to a drastic decrease in K with increasing I, as shown in Chap. 3. The protein zone then moves until it reaches the outlet of the column with the velocity close to the maximum velocity attained at I_R, the ionic strength of the elution buffer just as the peak of the protein zone emerges from the column. This situation is depicted in Fig. 2.13.

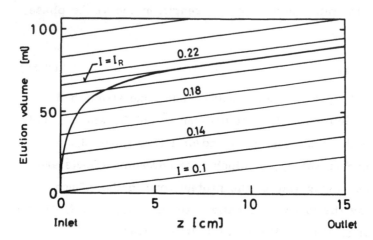

Fig. 2.13 Schematic diagram of the movement of the peak position in the column in linear gradient elution. Here, z is the distance from the inlet of the column. The thick line represents the trajectory of the peak position of the protein zone calculated by the equilibrium theory, which is described in detail in Sec. 2.2.3. The thin lines represent the isoionic strength lines. For example, the lowest line indicates that the ionic strength zone with I = 0.1 M (calculated from the salt concentration) moves with this line. Also, I_R is the ionic strength of the elution buffer at the exit of the column just when the peak emerges from the column. This distribution coefficient for the protein is considered dependent only on the ionic strength, and the experimental values for ovalbumin on DEAE-Sepharose CL6B (pH 7.9) shown in Chap. 8 were employed for the calculation. The values for the other experimental conditions employed are as follows. The column is d_c = 1.5 cm and Z = 15 cm; the slope of linear gradient g is 0.01 M/ml. The distribution coefficient for salt K' is 0.9; H = $(V_t - V_0)/V_0$ is 1.86 (see the text and the list of symbols for the above notations). (From Yamamoto et al., 1983a.)

As the protein zone moves down the column, it spreads because of the zone spreading effect described in Sec. 2.1. On the other hand, as shown in Fig. 2.14, dz/dt of the front part of the zone is always smaller than dz_p/dt since the ionic strength of the front part is lower than that of the peak. This causes flux opposite to that caused by zone spreading effects when we employ a moving frame with the peak position. Consequently, the front part of the protein zone is sharpened. Similarly, the rear part of the zone is also sharpened. When the protein concentration is so high that the protein follows a nonlinear concave isotherm, the protein zone has a long tail, as shown in Fig. 2.4. However, in linear gradient elution, the

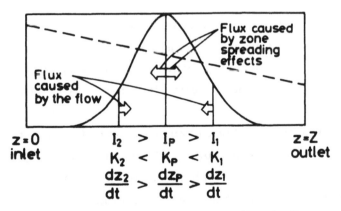

Fig. 2.14 Mechanism of zone spreading in linear gradient elution. The solid line represents the concentration distribution of proteins (the protein zone) in the column, which moves from the left to right of the figure. The broken line represents a linear gradient of the ionic strength in the column. Suffix p denotes the values at the peak positions of the zone. Suffixes 1 and 2 denote arbitrary positions on the front side and the rear side of z_p, respectively. Arrows in the figure show the directions of flux relative to the peak position. (From Yamamoto et al., 1983a.)

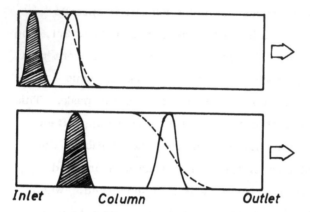

Inlet   *Column*   *Outlet*

Fig. 2.15 Elution mechanism in stepwise elution. Broken lines
indicate the spreading front boundary of the elution buffer.
Shaded areas are the protein zones that are partially desorbed
in the elution buffer; those not shaded are desorbed completely
in the buffer.

tail is also reduced by this zone sharpening effect since the nonlinearity of the isotherm usually decreases with an increase in I,
as shown in Chap. 3. The zone sharpening effect is one of the
characteristics of linear gradient elution that make this method
very attractive.

Figure 2.15 demonstrates the elution behavior of the protein
zone in the column in stepwise elution. When the ionic strength
of the elution buffer is high enough for a protein to be completely desorbed from the ion exchanger, the protein zone always locates in the spreading front boundary of the elution buffer, as shown by the unshaded area in Fig. 2.15. This situation
can be considered the gradient elution of the steep gradient, and
therefore zone sharpening as well as zone spreading occurs
during the elution.

On the other hand, when the protein is desorbed only partially at that ionic strength, the moving rate of any part of the

protein zone is constant or is governed by the protein concentration only and is lower than that of the elution buffer, as shown in the shaded area in the figure. In this case, the zone sharpening effect does not exist and the elution mechanism is similar to that of isocratic elution.

In order to predict the elution characteristics quantitatively, the zone sharpening effect caused by the time-dependent local moving rate of the protein zone mentioned earlier must also be taken into consideration.

2.2.2 Mathematical Model Based on the Plate Theory

A mathematical model is proposed for the elution of proteins on ion-exchange columns by a linear gradient increase and stepwise increase in ionic strength in order to predict relations between the elution characteristics (including the peak position and the peak width) and the operating conditions (the flow rate, the slope of the gradient, and others). This model is in principle based on the continuous-flow plate theory. Therefore, zone spreading effects are represented by the number of plates. Zone sharpening effects are also incorporated in the model by considering a change in ionic strength in each plate. The change in ionic strength, that is, the change in salt concentration during elution processes, is also predictable by plate theory. However, as shown later, the number of plates for salts is different from that for proteins since salts are much smaller molecules than proteins.

Assumptions

Since the present model is based on the continuous-flow plate theory described in Sec. 2.1.4, the following usual assumptions may be made.

1. The column consists of a certain number of equivalent theo-
    retical plates, in each of which the ratio of the volume of
    the stationary phase to that of the mobile phase is the same.
2. The flow is continuous without mixing between the plates.
3. The equilibrium of solutes between the two phases is in-
    stantaneously attained.

New assumptions made in this model are as follows:

4. The distribution coefficient K' of a salt between the mobile
    and stationary phases is dependent on neither concentration
    nor ionic strength and is not affected by the presence of
    the protein. Actually, these assumptions are valid in most
    cases, as shown experimentally in Chap. 8.
5. The number of plates for the protein N_p and that for the
    salt N_p' are constant.

The method for determining the number of plates is one of the
characteristics of the present model and is described in detail
in a later subsection.

Basic Equations for Protein and
Ionic Strength

Under assumptions 1, 2, and 5, a mass balance equation for
the protein in the nth plate is represented by Eq. (2.58) (see
Fig. 2.16):

$$F[C_{(n-1)} - C_{(n)}] = \frac{V_0}{N_p} \frac{dC_{(n)}}{dt} + \frac{V_t - V_0}{N_p} \frac{d\bar{C}_{(n)}}{dt} \qquad (2.58)$$

where $C_{(n)}, \bar{C}_{(n)}$ = protein concentrations in the mobile and
stationary phases at plate n, respectively

F = flow rate

t = time

V_0 = total volume of the mobile phase (the void volume of the
column)

V_t = total column volume

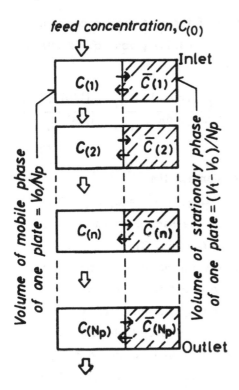

Fig. 2.16 Schematic drawing of a column according to the continuous-flow plate model. Note that for a mass balance equation for slat, n, N_p, $C_{(n)}$, and $\bar{C}_{(n)}$ should be read as n', N_p', $I_{(n')}$, and $\bar{I}_{(n')}$, respectively.

N_p = total number of theoretical plates for the protein in the column

$C_{(0)}$ = feed concentration

The first and second terms on the left-hand side imply the rate of the protein entering and leaving the nth plate with the fluid flow, respectively. The first term on the right-hand side describes the accumulation of the protein concentration in the mobile phase of the nth plate and the second, that in the stationary phase of the nth plate.

Under assumption 3, insertion of the distribution coefficient represented by Eq. (2.56) into Eq. (2.58) gives Eq. (2.59) (see Appendix B for this derivation):

$$\frac{dC_{(n)}}{d\theta} = \frac{N_p[C_{(n-1)} - C_{(n)}] - C_{(n)}(H)(dK\,[C_{(n)},\,I]/dI)(dI/d\theta)}{1 + H(K[C_{(n)},\,I] + C_{(n)}\,dK\,[C_{(n)},\,I]/dC_{(n)})}$$

$$(2.59)$$

where $\theta = tu/Z = Ft/V_0$ = dimensionless time.

The initial condition is $C_{(1)} = C_{(2)} = \cdots = C_{(N_p)} = 0$ at $\theta = -X_0$, where X_0 is the ratio of the sample volume to V_0. The boundary condition is $C_{(0)} = C_0$ for $-X_0 < \theta \leq 0$ and $C_{(0)} = 0$ for $\theta > 0$ where C_0 is the initial concentration of the sample protein. This is the basic equation for the protein and is similar in form to that presented by Nelson et al. (1978). Since this equation is nonlinear with respect to both protein concentration and ionic strength, it cannot be solved analytically. As shown later, Eq. (2.59) is numerically integrated by using the analytical solution for ionic strength, and its derivative with respect to θ derived in the following sections.

In a similar manner, a basic equation for the ionic strength $I_{(n')}$ in the mobile phase at plate n' is derived as Eq. (2.60) under assumptions 1—5:

$$\frac{dI_{(n')}}{d\theta} = R'[I_{(n'-1)} - I_{(n')}]$$

$$(2.60)$$

where $R' = N'_p/(1 + HK')$. K' is the constant distribution coefficient for the salt, and N'_p is the number of plates for the salt in the column. Because of assumption 4, Eq. (2.60) is a linear differential equation and can be solved analytically with known initial and boundary conditions. The solutions are shown in the following sections.

Equations for a Change in Ionic Strength for
a Stepwise Elution

The initial condition is

$$I_{(1)} = I_{(2)} = \cdots = I_{(N'_p)} = I_0 \qquad (2.61)$$

where I_0 is the initial ionic strength with which the ion-exchange
column is equilibrated. The boundary condition is $I_{(0)} = I_{elu}$
for $\theta > 0$, where I_{elu} is the ionic strength of the elution buffer.
Equation (2.60) can be solved analytically as

$$I_{(n')} = I_{elu} + (I_0 - I_{elu})g(\theta, n') \qquad (2.62)$$

$$g(\theta, n') = \sum_1^{n'} g'(\theta, n') \qquad (2.63)$$

$$g'(\theta, n') = (\theta R')^{n'-1} \frac{\exp(-\theta R')}{(n' - 1)!} \qquad (2.64)$$

The details of the derivation of these equations are given in
Appendix C. Differentiation of Eq. (2.62) with respect to θ
gives $dI_{(n')}/d\theta$:

$$\frac{dI_{(n')}}{d\theta} = (I_{elu} - I_0)g'(\theta, n')R' \qquad (2.65)$$

Equations for a Change in Ionic Strength
for Linear Gradient Elution

A linear gradient increase in the ionic strength (salt concentration) can be produced conveniently with two cylindrical vessels, as shown in Fig. 6.4, or with an automatic gradient maker.
The linear gradient at the inlet of the column is described by
Eq. (2.66):

$$I = I_0 + gFt \qquad (2.66)$$

where the slope of the gradient g is defined as

$$g = \frac{I_f - I_0}{V_G} \tag{2.67}$$

where I_f and I_0 = ionic strengths of final and initial starting buffers, respectively. V_G is a gradient volume, the sum of the volume of the mixing vessel and the reservoir in a convenient two-vessel apparatus, shown in Fig. 6.4, or the product of the gradient time t_G and the flow rate F in an automatic gradient maker ($V_G = F\ t_G$).

The basic equation and the initial condition for a linear gradient are the same as Eqs. (2.60) and (2.61), respectively. The boundary condition is $I_{(0)} = I_0 + G\theta$ for $\theta > 0$, where $G = gV_0$. The analytical solution is described as

$$I_{(n')} = I_{(n'-1)} - \frac{G}{R'} + \frac{Gg(\theta,\ n')}{R'} \tag{2.68}$$

Equation (2.68) is differentiated with respect to θ and becomes

$$\frac{dI_{(n')}}{d\theta} = G\ [1 - g(\theta,n')] \tag{2.69}$$

Details of the derivation of Eqs. (2.68) and (2.69) are shown in Appendix C.

If the zone spreading effects for salts are negligible in the case of $N_p' \to \infty$ (in the ideal case), the ionic strength in the nth plate is described by the simple equation

$$I_{(n)} = \begin{cases} I_0 & \text{for } \theta \leq \dfrac{(1 + HK')n}{N_p} \\[4mm] I_0 + G\left[\theta - \dfrac{(1 + HK')n}{N_p}\right] & \text{for } \theta > \dfrac{(1 + HK')n}{N_p} \end{cases} \tag{2.70}$$

$dI_{(n)}/d\theta$ is given by

$$\frac{dI_{(n)}}{d\theta} = \begin{cases} 0 & \text{for } \theta \leq \dfrac{(1 + HK')n}{N_p} \\[3mm] G & \text{for } \theta > \dfrac{(1 + HK')n}{N_p} \end{cases} \qquad (2.71)$$

Mehtod for Determining the Number of Plates

We determine the total number of plates N_p on the basis of the HETP equation (2.46) described in Sec. 2.1.5, which includes the constant distribution coefficient K, the longitudinal dispersion coefficient D_L, and the gel-phase diffusion coefficient, $\overline{D}$ (see Sec. 2.1.5). Then, $N_p = Z/HETP$ is given as

$$N_p = \frac{Z}{2D_L/u + d_p^{\,2}HKu/[\,30\overline{D}(1 + HK)^2]} \qquad (2.72)$$

where   Z = column length

u = actual velocity of the solvent flow

d_p = diameter of the packing materials

The justification of Eq. (2.72) was demonstrated in Sec. 2.1.6.

As discussed in Sec. 2.1.5 and Chap. 5, it can be assumed that D_L/u is constant regardless of flow rate or the type of gel particles and solutes and is a unique function of the particle size of gels. D_L/u would thus remain constant despite a change in ionic strength. On the other hand, the distribution coefficient and gel-phase diffusion coefficient for proteins are usually functions of ionic strength, as shown in Chaps. 3 and 4. Thus N_p varies with time during the elution process. However, it is difficult to treat this time-dependent N_p mathematically.

During passage down through the greater part of the column, the protein zone is mainly subjected to the ionic strength near

I_R as shown in Fig. 2.13.   Therefore, we assume that the number of plates for proteins N_p is constant and that $\bar{D}$ and K are
given by the values at $I = I_R$, $\bar{D}_R$, and K_R. N_p is then given
by the equation

$$N_p = \frac{Z}{2D_L/u + d_p^2 HK_R u/[30\bar{D}_R(1 + HK_R)^2]} \qquad (2.73)$$

The parameters D_L/u, $\bar{D}$, and K in Eq. (2.73) can be obtained
by the experimental methods shown in Chaps. 3 through 5.
However, since both $\bar{D}$ and K depend on I, they must be measured as a function of I, which is rather laborious.   The following expedient method is thus proposed for the determination of
N_p.

Macroporous ion exchangers, such as agarose ion exchangers,
may be regarded as gel chromatographic media when the ionic
strength is so high that electrostatic interaction between the ion
exchangers and proteins is negligible.   When the moment method
is applied to pulse response experimental results carried out
under such conditions, the $\bar{D}$ and K values obtained are critical
values that do not change with further increase in I, although
they are usually functions of I.   Therefore, we add the suffix
crt to these parameters.   For the relation between these critical
values and the values at $I = I_R$, the following equation is theoretically derived when ion exchangers do not swell or shrink
markedly with the change in ionic strength (see Sec. 4.1):

$$K_{crt}\bar{D}_{crt} = K_R\bar{D}_R \qquad (2.74)$$

This relation means that the gel-phase diffusion coefficient $\bar{D}$
decreases with an increase in electrostatic interaction between

proteins and ion exchangers. Insertion of this relation into Eq. (2.73) yields

$$N_p = \frac{Z}{2D_L/u + d_p^2 uHK_R^2/[30\bar{D}_{crt}K_{crt}(1 + HK_R)^2]} \qquad (2.75)$$

This equation is useful since N_p can be calculated for a given set of operating conditions with known values of $\bar{D}_{crt}$, K_{crt}, D_L/u, and K_R. The method of determining K_R is described later.

Since the distribution coefficients for salts are assumed to be constant irrespective of concentration, as already mentioned, the number of paltes for salts N_p' is given by

$$N_p' = \frac{Z}{2D_L/u + d_p^2 HK'u/[30\bar{D}'(1 + HK')^2]} \qquad (2.76)$$

where $\bar{D}'$ is the gel-phase diffusion coefficient for salts. Since salts are much smaller molecules than proteins, $\bar{D}'$ is greater than $\bar{D}$ and consequently N_p' is greater than N_p.

Calculation Method for Elution Profiles
of Proteins and Salts

When the adsorption isotherms $K[C, I]$ of a protein as a function of ionic strength and the number of plates N_p and N_p', for the protein and that for a salt are known, Eq. (2.59) can be calculated numerically with Eqs. (2.62) and (2.65) for a step-wise elution and with Eqs. (2.68) and (2.69) or Eqs. (2.70) and (2.71) for a linear gradient elution. Numerical calculations of the elution profiles of the protein (elution curve) and the salt, that is, $C_{(N_p)}$ and $I_{(N_p')}$ as a function of θ, can be carried out with a computer by the Runge-Kutta-Gill method

(Lapidus, 1962). The details of numerical calculation procedures are shown in Appendix D. The results are shown in Chap. 8.

As previously mentioned, the value of K_R or I_R, as well as the parameters obtained by the moment method, is required to determine the N_p value. Note that I_R is readily obtained by measurement of the ionic strength of the peak position of experimental elution curves. Alternative methods are as follows. Tentatively calculated results showed that the peak position is not affected by N_p when the N_p value is larger than 10. This means that I_R is obtained by a numerical calculation of the elution curve with an arbitrary value of N_p greater than 10 for given experimental conditions. In linear gradient elution, a simple graphic method for predicting I_R, described in Sec. 2.2.3, is also applicable.

2.2.3 Simple Methods for Predicting the Peak Position and Peak Width in Linear Gradient Elution

The dependency of the peak position and peak width on the slope of the gradient has been reported (Novotny, 1971; Kawasaki and Bernardi, 1970; Kato et al., 1982a). Although the model in Sec. 2.2.2 can predict these dependencies, it is necessary to calculate the whole elution curve for each column and operating condition. Moreover, the calculation requires a high-speed, large-memory computer. Simpler methods are thus desirable. We propose simple methods for these dependencies. The assumptions common to these methods are as follows.

a. The distribution coefficient K is not protein concentration dependent but a unique function of ionic strength; that is, $\bar{C} = K[I]C$.

b. The linear gradient of ionic strength in the column is ideally established as described by Eq. (2.70).

Graphic Method for Prediction of the Peak
Position in Linear Gradient Elution

When a protein solution is assumed to be adsorbed as an in-
finitesimally thin layer at the top of the column in addition to
assumptions a and b, the moving velocity of the protein zone
in a linear gradient elution is described by the equilibrium
theory (DeValut, 1943; Weiss, 1943):

$$\frac{dz_p}{dt} = \frac{u}{1 + HK[I]} \tag{2.77}$$

The ionic strength (salt concentration) at z_p is given as a function of time by

$$I = I_0 + G \, \frac{ut - z_p(1 + HK')}{Z} \tag{2.78}$$

When Eqs. (2.77) and (2.78) are solved simultaneously, we obtain the time when the peak of an elution curve appears in the
linear gradient elution. This concept was proposed by Drake
(1955) and also by Freiling (1955). However, the analytical
solution can be obtained only for some simplified cases in which
K' = 0 (a gradient substance is not accessible to the stationary
phase) and K[I] is given by a simple function of I (Drake, 1955;
Jandera and Churacek, 1974a; Morris and Morris, 1964). Otherwise, it is necessary to numerically integrate Eqs. (2.77) and
(2.78) with respect to time, and this integration must therefore be carried out for each operating condition.

On the other hand, we proposed a useful graphic method
for predicting the peak position in a linear gradient elution
(Yamamoto et al., 1983a). Our procedure consists of the integration of Eqs. (2.77) and (2.78) with respect to I instead of
time, as follows. We first differentiate Eq. (2.78) with respect
to z_p to obtain

$$\frac{dI}{dz_p} = G \frac{[u(dt/dz_p) - (1 + HK')]}{Z} \tag{2.79}$$

Insertion of Eq. (2.77) into the right-hand side of this equation gives

$$\frac{dI}{dz_p} = \frac{GH(K - K')}{Z} \tag{2.80}$$

The equation is integrated and gives

$$\int_{I_0}^{I_R} \frac{dI}{K - K'} = GH \int_0^Z \frac{1}{Z} \, dz_p$$

$$= GH \tag{2.81}$$

Let us define a function h(I) given by the equation

$$h(I) = \int_{I_a}^I \frac{dI}{K - K'} \tag{2.82}$$

where I_a = arbitrary ionic strength lower than any initial ionic strength I_0 for chromatography. Using this function, Eq. (2.81) can be represented by

$$h(I_R) - h(I_0) = GH \tag{2.83}$$

When $(K - K')$ values are known as a function of I, h(I) is calculated numerically or analytically for any arbitrary value of I and is plotted against I. Once the chart h(I) versus I is prepared according to this procedure, the I_R value is readily predictable from Eq. (2.83) using four operating variables: the slope of the gradient, the initial ionic strength, the total column volume, and the void volume of the column. The time at which I_R reaches the outlet of the column θ_R is then calculated by

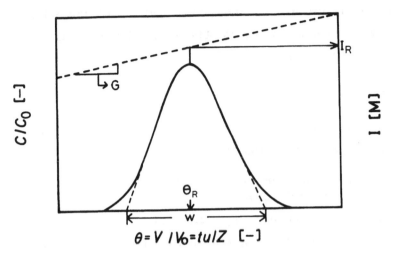

Fig. 2.17 Elution curve in linear gradient elution. The solid curve is the elution profile of a protein at the outlet of the column (elution curve), which has its maximum at $\theta = \theta_R$. The broken line is the linear gradient (salt concentration) at the outlet of the column. I_R is the ionic strength of the peak of the elution curve; that is, $I_R = I$ at $\theta = \theta_R$.

substituting I_R in a relation between I and θ at $z = Z$, shown by the following equation (see Fig. 2.17):

$$\theta_R = \frac{I_R - I_0}{G} + 1 + HK'$$ (2.84)

In addition, this method can also serve as (1) a shortcut method for prediction of the peak position in a linear gradient elution when K[I] is unknown and (2) a method for the determination of K as a function of I (Yamamoto et al., 1987d). The former method is as follows. We first measure I_R for various g with a given column and then plot them against GH. This plot can be employed for the prediction of the peak position of various column dimensions and the slope of the gradient if pH and ion exchanger are the same.

This plot is also used in the second method with the aid of
the following equation, which is obtained by differentiation of
Eq. (2.81) with respect to I:

$$\frac{d(GH)}{dI} = \frac{1}{K[I] - K'}$$
(2.85)

This equation implies that differentiation of the plot with respect
to I gives $1/(K - K')$ at that I.

Asymptotic Equation for Predicting the Peak Width in a Linear Gradient Elution

As mentioned previously, Eq. (2.59) cannot be solved analytically. However, an asymptotic solution that predicts the
upper limit of peak width in a linear gradient elution can be
derived from a quasi-steady-state model. As illustrated in Fig.
2.14, the zone spreading effects caused by the longitudinal dispersion and the gel-phase diffusion broaden the width of the
protein zone and the zone compression effects caused by the flow
tend to sharpen it during the linear gradient elution process.
During the early period of elution, the zone spreading effects
are dominant and the width of the protein zone increases. However, the zone compression effects increase with an increase in
the zone width since the contribution of the zone compression
effects caused by the flow increases with an increase in the
distance of the peak position of the zone. Finally, zone compression (sharpening) effects may balance zone spreading effects
and subsequently the width of the protein zone increases no
more; that is, a quasi-steady state is attained. An analytical
solution for the peak width at the quasi-steady state is derived
from Eq. (2.59) by the procedure shown in Appendix E. In
addition to assumptions a and b, we assume for derivation of

the solution that the protein zone moves with a velocity of
$u/(1 + HK_R)$. This assumption may be reasonable since the
peak position is subjected to the ionic strength near I_R during
the later period of elution, as shown in Fig. 2.13. The variance
of the elution curve σ_θ^2 (see Fig. 2.17) thus obtained is expressed in the rearranged form

$$\frac{\sigma_\theta^2 N_p}{(1 + HK_R)^2} = -\frac{1}{2} \frac{1 + HK_R}{GH(1 + HK')(dK/dI)} \tag{2.86}$$

Another asymptotic solution at extremely low G values is derived
as Eq. (2.87) (Jandera and Churacek, 1974a):

$$\frac{\sigma_\theta^2 N_p}{(1 + HK_R)^2} = 1.0 \tag{2.87}$$

This equation implies that the peak width in the linear gradient
elution is equal to that of the isocratic elution of $I = I_R$.

2.2.4 R_s in Linear Gradient Elution

As mentioned in Sec. 2.1.6, the separation efficiency of two
components is governed by two factors: the distance between
two adjacent peaks and their widths. A large number of variables affect separation in linear gradient elution, as described
earlier. Some of them may affect only one of the two factors,
but others can contribute to both factors. Therefore, if these
variables can be related to the R_s equation defined by Eq.
(2.51), the equation will be very helpful for the prediction and
optimization of the separation.

We therefore try to derive an R_s equation in a linear gradient
elution by using an asymptotic solution from a quasi-steady-

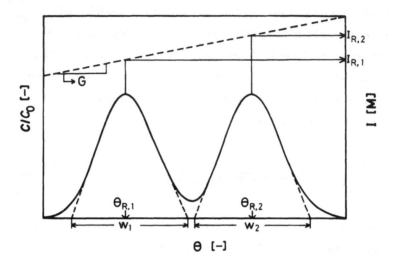

Fig. 2.18 R_s in linear gradient elution. The meanings of symbols and lines are the same as those in Fig. 2.17.

state model, Eq. (2.86). Insertion of Eq. (2.84) into Eq. (2.52) gives (see Fig. 2.18)

$$R_s = \frac{I_{R,2} - I_{R,1}}{2G(\sigma_{\theta,1} + \sigma_{\theta,2})} \tag{2.88}$$

We assume that $\sigma_{\theta,1}$ is equal to $\sigma_{\theta,2}$ and can be given by Eq. (2.86) in Sec. 2.2.3, and $K' = K_{I_R}$. Then, insertion of Eq. (2.86) into Eq. (2.88) yields

$$R_s = \frac{(I_{R,2} - I_{R,1})[N_{p,1}(-dK_1/dI)_{I=I_{R,1}} H]^{1/2}}{(1 + HK_{I_{R,1}})(8G)^{1/2}} \tag{2.89}$$

It is further assumed that Eq. (2.75) is valid for $N_{p,1}$. Equation (2.75) is rewritten as

$$N_{p,1} = \frac{Z}{(A + C_1 u)} \tag{2.90}$$

where

$$A = \frac{2D_L}{u} \tag{2.91a}$$

and

$$C_1 = \frac{d_p^2 H K_{R,1}^2}{30(1 + H K_{R,1})^2 \bar{D}_{crt,1} K_{crt,1}} \tag{2.91b}$$

[For the meanings of the variables in Eq. (2.91), see Sec. 2.2.2.] From Eqs. (2.89) and (2.90), the following R_s equation is then obtained:

$$R_s = \frac{(I_{R,2} - I_{R,1})[Z(-dK_1/dI)_{I=I_{R,1}} H]^{1/2}}{(1 + H K_{R,1})[8gV_0(A + C_1 u)]^{1/2}} \tag{2.92}$$

From this equation, the following interesting findings are made:

1. R_s increases with the decrease in the slope of the gradient g: $R_s \propto g^{-1/2}$.
2. When the slope of the gradient g is constant. R_s is not dependent on the column length Z.
3. Under the usual conditions, HETP increases with flow rate. So R_s is improved by a decrease in flow rate. Furthermore, if the N_p equation (2.75) is valid, R_s^{-2} becomes a linear function of u. When the flow rate is high so that A is negligible compared with $C_1 u$ in Eq. (2.92), R_s is proportional to $u^{-1/2}$.
4. When G (= gV_0) is constant (that is, g should be reduced in inversely proportion to column length), R_s increases with the column length: $R_s \propto Z^{1/2}$.

The validity of these findings is discussed in Sec. 8.3, where several experimentally measured R_s and numerically calculated

R_s are correlated with some of the variables that appear in Eq. (2.92).

2.2.5 Criticism of the Model and Previous Equations (Secs. 2.2.2 Through 2.2.4) in Comparison with Other Models or Equations

The chief advantages of the model described in Sec. 2.2.2, developed from the plate theory, are its simplicity and wide applicability. The elution properties of proteins with a concave or convex exponential gradient increase in ionic strength could also be predicted by the model, as well as those with a linear gradient increase. In spite of these advantages, this model is not applicable to cases in which the effect of mass transport is significant, that is, highly asymmetric elution curves due to low gel-phase diffusion coefficients.

Several different methods are available for determining the number of plates. A commonly employed method is to measure the peak width obtained from the isocratic elution experiment with a small sample (pulse input). N_p is then calculated by Eq. (2.37) or (2.44). N_p can also be determined from the breakthrough curve (Reilley et al., 1962). However, these standard methods are not directly applicable to the model we described in Sec. 2.2.2. Nelson et al. (1978) presented a model for cyclic separation processes that is similar to our model. In their model, the number of plates was determined by trail-and-error matching of theoretical with experimental results. This procedure is troublesome and time consuming, however. Moreover, in their model the values of N_p and N_p' are considered equal. The method adopted for determining N_p and N_p' in our model is based on the phenomenological parameters related to the mass transport of a solute in the column. Once $\overline{D}_{crt}$ and K_{crt}

for a given protein and $\overline{D}'$ and K' for a given salt are determined
for a given ion exchanger, N_p and N'_p are determined uniquely
for arbitrary operating conditions by Eqs. (2.75) and (2.76),
respectively. However, as previously mentioned, Eq. (2.75)
may not be applicable to ion exchangers that swell or shrink
markedly with the change in the ionic strength since Eq. (2.74)
is not valid for such ion exchangers. Although Fig. 2.13 suggests justification of the assumptions made for the method for
determining the number of plates, their applicable range must
be judged by comparing experimental elution curves with the
theoretical curves shown in Chap. 8.

Several workers have reported the influence of the slope of
the gradient on the ionic strength at which the peak is eluted
(Novotny, 1971; Kawasaki and Bernardi, 1970; Kato et al., 1982a).
A graphic method for determining I_R is developed on the basis
of the same concepts as both Freiling's (1955) and Drake's techniques (1955). Our method is more convenient than their methods, since the graph, once made for a given protein and ion
exchanger, can be used to determine I_R for various operating
conditions. However, this method is not directly applicable
when the protein concentration is so high that the distribution
coefficient is dependent on it as well as on the ionic strength.
Unfortunately, no simple method is available for this case.
Drake (1955) presented an approximate graphic method by which
one can expect how the peak retention time changes with the
solute concentration qualitatively. Since the moving rate is
governed by both the protein concentration and the ionic
strength, the change in concentration due to zone spreading
must also be taken into consideration for prediction of the peak
position. This can be done only by a model like that described

in Sec. 2.2.2. As mentioned in Sec. 2.2.3, Eq. (2.81) can also work as a shortcut method for the prediction of I_R and for the determination of K when K is unknown. We also should stress here that Eq. (2.81) is the only exact equation that relates K from isocratic elution to the peak position (retention time) in linear gradient elution, regardless of the salt concentration (ionic strength) dependence of K. The analytical solution can be obtained in some simplified cases (Drake, 1955; Jandera and Churacek, 1974a; Morris and Morris, 1964; Chang et al., 1980). In these studies, K is assumed to have an exponential dependence or a power-law dependence on I. Chang et al. (1980) have derived such solutions for a column saturated with a sample before elution as well as a pulse input by using the exponential dependence on I of K. The solution when K is a power-law function of I is given in Sec. 8.2.

Dependence of the peak width on the slope of the gradient was also reported (Novotny, 1971; Kawasaki and Bernardi, 1970) and explained qualitatively (Drake, 1955; Freiling, 1955; Morris and Morris, 1964; Peterson, 1970; Kawasaki and Bernardi, 1970). We solved a quasi-steady-state model and obtained an asymptotic solution for the peak width in a linear gradient elution. The concept of the quasi-steady state was first presented by Svenson (1961) for isoelectric focusing and then employed by Sluyterman and Elgersma (1978) for chromatographic focusing. The quasi-steady state may be attained when the slope of the gradient is steep enough or the column is long enough, as shown by Drake (1955). However, no theoretical basis was established for the range of applicability of the quasi-steady-state model. In Chap. 8, these simpler methods, as well as the model developed from the plate theory, are compared with experimental results in order to show their validity and range of applicability.

R_S is very useful when it is related to variables influencing separation. By using such an R_S equation, we can predict R_S before operation when the values of phenomenological parameters in the equation are known. Even when the values of such parameters are unknown, the separation can be readily optimized from a small amount of experimental data. However, an R_S equation can be obtained as a function of operating and column variables only in a limited case, such as linear chromatography in which the distribution coefficient is constant and its elution is isocratic. In this case, an R_S equation is given by Eq. (2.55) in Sec. 2.1.4. This equation gives us some important information, including the well-known relation that R_S increases in proportion to the square root of the column length. In addition, with an HETP equation suitable for the system considered, the dependence of R_S on flow rate may also be predictable. Unfortunately, no exact analytical R_S equation for linear gradient elution IEC of proteins is available. Several interesting findings were extracted from the R_S equation derived in Sec. 2.2.4. For example, the effect of column length on R_S is not as simple as that in isocratic elution chromatography in which the distribution coefficient is constant, as described in Sec. 2.1.7. However, since this equation has been derived with several assumptions, its validity must be judged by examining how well it predicts the experimental results. This is demonstrated in Sec. 8.3.

Snyder and his colleagues (Snyder et al., 1983; Stout et al., 1986; Stadalius et al., 1987) developed a model for predicitng the peak retention time t_R and the peak width W in linear gradient elution reversed-phase chromatography of proteins and applied it to the IEC of proteins. They also presented the relations between R_S and the operating variables, which are similar to those given in Sec. 2.2.4.

3

Ion-Exchange Equilibria

Equilibria between ion exchangers (ion-exchange resins) and solutions are one of the most important factors in ion-exchange chromatography. The equilibrium relationship determines the distribution of solutes (proteins) between gel phase (stationary phase) and outer solution (mobile phase). It is influenced by various factors, such as ionic strength of the solution, protein concentration, pH, and types of ion exchangers. The dependency of the distribution coefficient of protein on ionic strength and pH is closely related to the elution pattern, that is, peak position, peak width, and so on (see Chap. 2). Thus, equilibrium should be expressed in a quantitative fashion as a function of the system variables in order to predict the elution profile in column operations. However, experimental as well as theoretical approaches to equilibria with proteins are considerably hindered in comparison with those for low-molecular-weight inorganic ions (e.g., Helfferich, 1962), since the ion-exchange chromatography of proteins is a relatively new technology and equilibria with macromolecules, such as proteins, are much more

complicated than those with inorganic ions. Special regard
should be paid to the properties of proteins. Although proteins
are biopolymers composed of distinct monomer units, that is,
amino acids linked by peptide bonds, their physical properties
are usually much different from those of the simple assembly of
individual amino acids. In an extreme case, those proteins with
the same net charge and molecular weight but with different
configurations exhibit different behaviors against ion exchangers.
Various interactions, such as hydrophobic interaction and hydrogen bonding, in addition to electrostatic forces, exist among
protein molecules themselves and between proteins and fixed
ionic groups on the matrix, or small ions. Furthermore, proteins are not as stable as inorganic ions. They are apt to be
denatured in acidic or alkaline solution. Many enzyme proteins
are denatured even at room temperature. Denaturation causes
conformational changes in a protein; that is, a native protein
with closely packed structure becomes randomly coiled or co-
agulated on denaturation. Ion-exchange experiments with proteins are thus troublesome and considerably difficult to perform.

3.1 THEORETICAL TREATMENT

Equilibria between ion exchangers and protein solutions may
be essentially described by means of rigorous thermodynamics,
with the fixed ionic groups, the dissolved electrolytes, and the
solvent as the components. Although rigorous thermodynamic
treatment is correct and universal, its practical value is considerably restricted owing to the mathematical complexity of its
many variables. Therefore, models that can reasonably explain
the phenomenon of ion-exchange equilibria are required practically. As in equilibria with inorganic ions (e.g., Helfferich,

1962), there are in principle three approaches for expression of
the equilibrium relationship, that is, those based on (1) the law
of mass action, (2) the Donnan potential, and (3) the empirical
or semiempirical equation.

In this section, we briefly introduce these models in connec-
tion with ion-exchange equilibria of proteins.

3.1.1 Approach Based on the Law of Mass Action

According to the law of mass action, an ion-exchange reaction is
expressed by Eq. (3.1), assuming that uniform bonds are
formed between an ion exchanger with fixed ionic groups and
an oppositely charged protein with an electrochemical valence Z_P,
with the concomitant displacement of the counterion with valence
Z_B; the effect of coions and proton is negligible.

$$P + \left(\frac{Z_P}{Z_B}\right) \overline{B} \;\rightleftharpoons\; \overline{P} + \left(\frac{Z_P}{Z_B}\right) B \tag{3.1}$$

where   P, B = protein and counterion in the solution

$\overline{P}$, $\overline{B}$ = protein and counterion in the ion exchanger-bound state

K_e = equilibrium constant for Eq. (3.1)

The overbar indicates the ion exchanger-bound state.

The equilibrium constant K_e is represented as

$$K_e = \frac{(\overline{a}_P)^{|Z_B|}\,(a_B)^{|Z_P|}}{(a_P)^{|Z_B|}\,(\overline{a}_B)^{|Z_P|}} \tag{3.2}$$

When counterion B is monovalent, Eq. (3.2) becomes

$$K_e = \frac{\overline{a}_P}{a_P}\left(\frac{a_B}{\overline{a}_B}\right)^{|Z_P|} \tag{3.3}$$

In Eq. (3.3), a denotes the activity of the species. Using the activity coefficient, Eq. (3.3) is expressed in the form of molarity as

$$K_e = \frac{\overline{m}_P}{m_P} \left(\frac{m_B}{\overline{m}_B} \right)^{|Z_P|} \frac{\overline{\gamma}_P}{\gamma_P} \left(\frac{\gamma_B}{\overline{\gamma}_B} \right)^{|Z_P|} \qquad (3.4)$$

and

$$\gamma_i = \frac{a_i}{m_i} \qquad i = B, P \qquad (3.5)$$

where m and γ = molarity and activity coefficients of the species, respectively.

The electroneutrality condition in the ion exchanger (resin) is

$$\overline{m}_R = |Z_P| \overline{m}_P + \overline{m}_B \qquad (3.6)$$

where m_R is the molarity of the fixed ionic groups on the ion exchanger.

For practical purposes, ion-exchange equilibria should be expressed in the form of the distribution coefficient K, defined as the ratio of the molarity of the resin-bound protein to that in the solution; that is, $K = \overline{m}_P/m_P$. K is obtained from Eqs. (3.4) and (3.6) as

$$K = \frac{\overline{m}_P}{m_P} = K_e \Gamma_1 \left(\frac{\overline{m}_R - |Z_P| \overline{m}_P}{m_B} \right)^{|Z_P|} \qquad (3.7)$$

where

$$\Gamma_1 = \left(\frac{\overline{\gamma_B}}{\gamma_B}\right)^{|Z_P|} \frac{\gamma_P}{\overline{\gamma_P}} \tag{3.8}$$

When $z_P m_P < m_R$, Eq. (3.7) reduces to

$$K = \frac{(K_e \Gamma_1)(m_R/m_B)^{|Z_P|}}{1 + (K_e \Gamma_1)(m_R/m_B)^{|Z_P|} Z_P^2 m_P/m_R} \tag{3.9}$$

Equation (3.9) indicates that at the fixed pH, the K value depends on the molarities of the fixed ionic groups on the ion exchanger, the inorganic ion and protein in the solution, and the
valence of the protein. The K value diminishes as the values
of m_B, m_P, and Z_P increase and the m_R value decreases. When
$\overline{m}_P$ is plotted against m_P at fixed values of m_B, m_R, and Z_P,
one can obtain a typical saturation curve as is often found in
the experimental results (see Fig. 3.1). In particular, if m_P is
very much smaller than m_R, that is, m_B is virtually unchanged
by the adsorption of proteins, Eq. (3.9) becomes

$$K = K_e \Gamma_1 \left(\frac{m_R}{m_B}\right)^{|Z_P|} \tag{3.10}$$

The K value expressed by Eq. (3.10) is independent of m_P but
strongly depends on the concentration and/or ionic strength of
the outer solution (m_B). When there are other types of interaction, such as hydrogen bonding, hydrophobic interaction, in
addition to electrostatic interaction, Eq. (3.10) take a different
form. The exponent on the right-hand side of Eq. (3.10) may
also be deviated from Z_P.

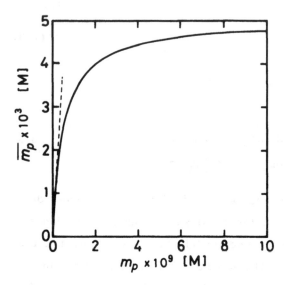

Fig. 3.1 Schematic representation of ion-exchange equilibria of a protein according to Eq. (3.9) ($K_e \Gamma_1 = 1.0$, $m_R/m_B = 5$, $Z_P = 10$, $m_R = 0.5$ M).

The approach based on the law of mass action is quite simple and has proven to be coincident at least qualitatively with experimental results obtained with ion-exchange resins, as shown later. However, this model does not account for the effect of degree of cross-linking and the protein size, since the swelling equilibria are ignored. It is also difficult to explain the pH difference between ion exchanger and outer solution or the effect of coions.

3.1.2 Approach Based on the Donnan Potential

This approach has been developed principally in ion-exchange reaction of inorganic ions with various ion-exchange resins. A number of theoretical equations have been proposed to express quantitatively the equilibrium relation based on the Donnan poten-

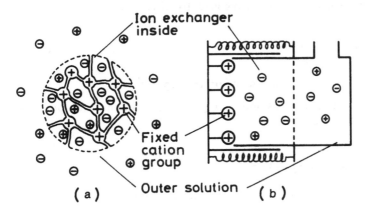

Fig. 3.2 Actual anion-exchange resin (or cation exchanger) (a) and Gregor's model of an elastic matrix (b). In the case of an cation-exchange resin (or cation exchanger), the fixed ionic groups show a minus charge.

tial (Gregor, 1948, 1951; Glueckauf and Duncan, 1951, 1952; Lazare et al., 1956; Stone, 1962; and others). In the simplest of these treatments, the model of the elastic matrix first developed by Gregor is used. Gregor considered the matrix of the ion-exchange resin a network of elastic springs (see Fig. 3.2). When resin swells, the force opposing the spring tension arises, yielding the osmotic pressure difference between the gel phase (ion exchanger and ion-exchange resin) and the outer solution phase.

In equilibrium, the electrochemical potential η_i of the ionic species i must be the same in both phases:

$$(\eta_i)^P = (\overline{\eta_i})^{\overline{P}} \tag{3.11}$$

where the overbar denotes the gel phase (the ion-exchange resin or ion exchanger), as shown previously. The electrochemical potential at pressure P is expressed by Eq. (3.12).

$$(\eta_i)^P = (\mu_i)^P + Z_i \mathfrak{F} \phi \tag{3.12}$$

where μ_i = chemical potential of the species i

   Z_i = electrochemical valance (negative for anions and positive
       for cations)

   $\mathfrak{F}$ = Faraday constant

   ϕ = electrical potential

If the pressure dependence of the partial molar volume is ignored, the μ_i at pressure P is given as

$$(\mu_i)^P = (\mu_i)^{P_0} + (P - P_0)v_i \tag{3.13}$$

where v_i is the partial molar volume and P_0 is the standard
pressure (usually 1 atm).

From Eqs. (3.12) and (3.13), $(\eta_i)^P$ and $(\overline{\eta_i})^P$ are derived:

$$(\eta_i)^P = \mu_i^{P_0} + \mathfrak{R}T \ln a_i + Pv_i + Z_i \mathfrak{F} \phi \tag{3.14}$$

$$(\overline{\eta_i})^P = \mu_i^{P_0} + \mathfrak{R}T \ln \overline{a_i} + \overline{P}\,\overline{v_i} + Z_i \mathfrak{F} \overline{\phi} \tag{3.15}$$

The Donnan potential E_{Don}, which is the electrical potential
difference $\overline{\phi} - \phi$, is obtained from Eqs. (3.14) and (3.15) as
Eq. (3.16), assuming that there are no differences in Z_i and v
at the both phases.

$$E_{Don} = \overline{\phi} - \phi$$

$$= \frac{1}{Z_i \mathfrak{F}} \left(\mathfrak{R}T \ln \frac{a_i}{\overline{a_i}} - \Pi v_i \right) \tag{3.16}$$

where

$$\Pi = \overline{P} - P \quad \text{swelling pressure} \tag{3.17}$$

Equation (3.16) holds for any mobile ionic species present in the system. For example, let us consider the simple case in which the system is composed of a negatively charged protein with electrochemical valence Z_P and a 1:1 valent salt (A^+, B^-) with a monovalent anion exchanger (R^+). Equation (3.16) becomes for P, A^+, B^-, H^+, and OH^-, respectively,

$$E_{Don} = - \frac{1}{|Z_P|\mathcal{F}} \left(\mathcal{R} T \ln \frac{a_P}{\overline{a}_P} - \Pi v_P \right) \tag{3.18}$$

$$E_{Don} = \frac{1}{\mathcal{F}} \left(\mathcal{R} T \ln \frac{a_A}{\overline{a}_A} - \Pi v_A \right) \tag{3.19}$$

$$E_{Don} = - \frac{1}{\mathcal{F}} \left(\mathcal{R} T \ln \frac{a_B}{\overline{a}_B} - \Pi v_B \right) \tag{3.20}$$

$$E_{Don} = \frac{1}{\mathcal{F}} \left(\mathcal{R} T \ln \frac{a_H}{\overline{a}_H} - \Pi v_H \right) \tag{3.21}$$

$$E_{Don} = - \frac{1}{\mathcal{F}} \left(\mathcal{R} T \ln \frac{a_{OH}}{\overline{a}_{OH}} - \Pi v_{OH} \right) \tag{3.22}$$

Equations (3.18) through (3.20) are further rearranged using activity coefficients:

$$E_{Don} = - \frac{1}{|Z_P|\mathcal{F}} \left(\mathcal{R} T \ln \frac{m_P \gamma_P}{\overline{m}_P \overline{\gamma}_P} - \Pi v_P \right) \tag{3.18'}$$

$$E_{Don} = \frac{1}{\mathfrak{F}} \left(\mathfrak{R}T \ln \frac{m_A \gamma_A}{\overline{m}_A \overline{\gamma}_A} - \Pi v_A \right) \tag{3.19'}$$

$$E_{Don} = -\frac{1}{\mathfrak{F}} \left(\mathfrak{R}T \ln \frac{m_B \gamma_B}{\overline{m}_B \overline{\gamma}_B} - \Pi v_B \right) \tag{3.20'}$$

$$E_{Don} = \frac{1}{\mathfrak{F}} \left(\mathfrak{R}T \ln \frac{m_H \gamma_H}{\overline{m}_H \overline{\gamma}_H} - \Pi v_H \right) \tag{3.21'}$$

$$E_{Don} = -\frac{1}{\mathfrak{F}} \left(\mathfrak{R}T \ln \frac{m_{OH} \gamma_{OH}}{\overline{m}_{OH} \overline{\gamma}_{OH}} - \Pi v_{OH} \right) \tag{3.22'}$$

The electroneutrality conditions in the ion exchanger and in the solution are

$$m_R + \overline{m}_A + \overline{m}_H = |Z_P| \overline{m}_P + \overline{m}_B + \overline{m}_{OH} \tag{3.23}$$

$$m_A + m_H = |Z_P| m_P + m_B + m_{OH} \tag{3.24}$$

The following equations are derived from Eqs. (3.18') through (3.22'):

$$\frac{m_P \overline{m}_B^{|Z_P|}}{\overline{m}_P m_B^{|Z_P|}} = \exp\left[\frac{\Pi(v_P - v_B|z_P|)}{\mathfrak{R}T} \right] \frac{1}{\Gamma_1} \tag{3.25}$$

$$\frac{m_A m_B}{\overline{m}_A \overline{m}_B} = \exp\left[\frac{\Pi(v_A + v_B)}{\mathfrak{R}T} \right] \frac{1}{\Gamma_2} \tag{3.26}$$

$$\frac{m_B m_H}{\overline{m}_B \overline{m}_H} = \exp\left[\frac{\Pi(v_B + v_H)}{\mathfrak{R}T} \right] \frac{1}{\Gamma_3} \tag{3.27}$$

$$\frac{m_{OH} m_A}{\overline{m}_{OH} \overline{m}_A} = \exp\left[\frac{\Pi(v_{OH} + v_A)}{\mathfrak{R}T} \right] \frac{1}{\Gamma_4} \tag{3.28}$$

where

$$\Gamma_1 = \frac{\gamma_P}{\bar{\gamma}_P}\left(\frac{\bar{\gamma}_B}{\gamma_B}\right)^{|Z_P|} \tag{3.29}$$

$$\Gamma_2 = \frac{\gamma_A \gamma_B}{\bar{\gamma}_A \bar{\gamma}_B} \tag{3.30}$$

$$\Gamma_3 = \frac{\gamma_B \gamma_H}{\bar{\gamma}_B \bar{\gamma}_H} \tag{3.31}$$

$$\Gamma_4 = \frac{\gamma_{OH} \gamma_A}{\bar{\gamma}_{OH} \bar{\gamma}_A} \tag{3.32}$$

Solving Eqs. (3.23) through (3.26) gives the distribution coefficient K as a function of protein concentration in the solution, ionic strength, and pH, when the information on swelling pressure, electrochemical valence of the protein, partial molar volumes, and activity coefficients of the solutes are available.

In the extreme case $\bar{m}_P \ll m_R$, one obtains the following equation for K, if there are no pH differences between both phases:

$$K = \left(\frac{m_R}{m_B}\right)^{|Z_P|} \exp\left[-\frac{\Pi}{\Re T}(v_P - v_B|Z_P|)\right] \Gamma_1 \tag{3.33}$$

Equation (3.33) is apparently coincident with Eq. (3.10), derived from the law of mass action. From Eqs. (3.10) and (3.33), the equilibrium constant K_e in Eq. (3.2) is expressed by $\exp[-\Pi(v_P - v_B|Z_P|)/\Re T]$. In other words, the equilibrium constant for the ion-exchange reaction is not constant but de-

pendent on the swelling pressure inside the gel, the partial
molar volume of the component, and so on. Moreover, the
swelling pressure Π is closely related to the degree of cross-
linking of the gel. Thus, the K value decreases as the concentration of the gel component increases even at the same
fixed ionic concentration of the ion exchanger and ionic strength.
The activity coefficient term Γ_1 implicity includes various factors other than the swelling pressure effect, such as ion-pair
formation, association, ionic solvation, and interactions between
coions. Several approaches were developed to explicitly express
these effects in the study of ion-exchange equilibria with inorganic ions (e.g., Helfferich, 1962). The important point to be
noted in the treatment of ion-exchange equilibria with proteins
is that the swelling pressure effect may become significant compared with those with inorganic ions, since the partial molar
volume of proteins is very large.

In the other extreme case in which the ionic strength of the
solution is very high so that $\overline{m}_B$ coincides with m_B, K is given
as

$$
\begin{aligned}
K &= \exp \left[- \frac{\Pi}{\mathcal{R}T} (v_P - v_B |Z_P|) \right] \Gamma_1 \\
&= \exp \left(- \frac{\Pi}{\mathcal{R}T} v_P \right) \Gamma_1
\end{aligned}
\tag{3.34}
$$

Under such conditions, the electrostatic interactions between
the fixed ions on the ion exchanger and protein or salt ions can
be ignored. Ginzburg and Cohen (1964) proposed an equation
similar to Eq. (3.34) in order to correlate the distribution coefficient of nonionic solutes with the partial molar volume of the
solute. On the other hand, using a statistical consideration,
Ogston (1958) derived the following equation for the distribution

coefficient of nonionic solutes for cross-linked dextran gels
(Sephadex):

$$K = \exp[-\pi L (r_s + r_r)^2] \tag{3.35}$$

where   L = concentration of rod (dextran molecule) expressed
              in cm rod per cm^3

r_s = equivalent radius of a solute molecule

r_r = radius of the rod

It is interesting that Eqs. (3.34) and (3.35) take a similar form,
although they were derived from completely different models.

3.1.3 Empirical or Semiempirical Equations

The most primitive but fairly practical approach is to develop
the empirical or semiempirical equation that fits the experimental
data, since the complete theoretical analysis of the ion-exchange
equilibria with proteins is extremely difficult, as seen in Secs.
3.1.1 and 3.1.2. Thus, a number of empirical or semiempirical
equations developed for equilibria with inorganic ions (e.g.,
Helfferich, 1962) may also be useful for proteins; these equations
are modifications of either the mass-action law or adsorption isotherms of the Langmuir or Freundlich type. Several proposed
equations are shown here.

The Wiegner-Jenny (1927) equation contains two empirical
constants:

$$y = k \left(\frac{C}{C_0 - C} \right)^{1/p} \tag{3.36}$$

where   y = counterions exchanged per unit weight of ion ex-
              changer

C = equilibrium concentration of exchanging ion in solution

C_0 = initial concentration of exchanging ion in solution

k, p = empirical constants

Vageler and Woltersdorf (1930a,b) proposed an equation similar to the Langmuir adsorption isotherm:

$$y = \overline{Q}\, \frac{Q}{Q + s} \tag{3.37}$$

where   Q = number of exchangeable counterions in solution

$\overline{Q}$ = weight capacity of the ion exchanger

s = empirical constant

The following equation for K with dilute concentrations of protein is derived from Eq. (3.10) or (3.33):

$$K = \frac{\overline{m}_P}{m_P} = A(I)^B \tag{3.38}$$

where I = ionic strength of the solution and A and B = empirical constants.

Yamamoto et al. (1983b) presented a modified equation for Eq. (3.38) that also takes into consideration the intake of proteins into the gel phase with a sieving effect:

$$K = \frac{\overline{m}_P}{m_P} = AI^B + K_{crt} \tag{3.39}$$

where K_{crt} is the critical distribution coefficient of the protein at high ionic strength, where electrostatic interaction can be ignored. As shown later, experimental K values at low protein concentrations obtained with various ion exchangers and proteins can be correlated well with Eq. (3.39).

3.2 EXPERIMENTAL METHODS

Ion-exchange equilibria (adsorption isotherm) between proteins
and ion exchangers can be measured in most cases either by a
batch method or by frontal analysis in a column. The batch
method has the following merits (Aranyi and Boross, 1974): (1)
it requires much smaller amounts of protein and ion exchanger;
(2) it is relatively rapid; (3) the distribution coefficients can
be determined over a wide range of concentrations; and (4) the
results are not influenced by diffusion or nonequilibrium effects.
Thus, equilibrium data have been obtained mainly by this method, except when there is a relatively low distribution coefficient
of proteins. When the K value is lower than around $2.0-5.0$,
frontal analysis is more accurate than the batch method. Moreover, in the extreme case in which the ionic strength of the
elution buffer is so high that electrostatic interaction inside the
ion exchanger can be ignored, the moment method can be used.

In measurement by the batch method, an appropriate amount
of gel should first be sufficiently swollen and equilibrated by
repeated exchange of the buffer solution until the pH of the
outer solution coincides with the buffer pH. Then, a small
amount of the swollen gel is weighed after removing the liquid
among the interstices of the gel particle using a glass filter of
suitable pore size. The weighed gel is mixed with a protein
solution of known concentration dissolved in the same buffer.
For simplicity of data analysis, the amount of the protein solution should be sufficiently large that the change in pH and in
the concentrations of salt ions due to exchanging reactions can
be ignored. The mixture is allowed to equilibrate with shaking
at a prescribed temperature. Several hours are usually neces-

sary to attain complete equilibration. The concentration of
protein in the gel is easily calculated by the equation

$$\bar{m}_P = \frac{W_s}{W_g} (m_P^i - m_P^f)$$ (3.40)

where $\bar{m}_P$ = molarity of the protein in the ion exchanger

m_P^i, m_P^f = molarities of the protein in the outer solutions
at the initial and equilibrium states

W_s = amount of the protein solution

W_g = amount of the swollen gel

In frontal analysis (Gleuckauf et al., 1949), an appropriate
volume of swollen gel is packed in a column and the column is
equilibrated with the buffer solution. A protein, dissolved in
the same buffer as was used for equilibration, is applied to the
inlet of the column until the protein concentration at the exit
of the column becomes identical to that at the inlet. The protein concentration in the gel is calculated from the total amount
of the protein eluted Q_e, the amount of the protein applied Q_s,
the void volume of the column V_0, and the total column volume
V_t as

$$\bar{m}_P = \frac{Q_s - Q_e - V_0 \rho_s m_P^f}{(V_t - V_0) \rho_g}$$ (3.41)

where ρ_s and ρ_g are the density of the solution and that of the
swollen gel, respectively. m_P^f is the molarity of the protein
applied at the column inlet. Q_e can be calculated by integrating
the shaded area in Fig. 3.3.

When electrostatic interaction inside the ion exchanger can
be ignored, the pulse response method is useful. A pulse response experiment is carried out as follows (Mehta et al., 1973;

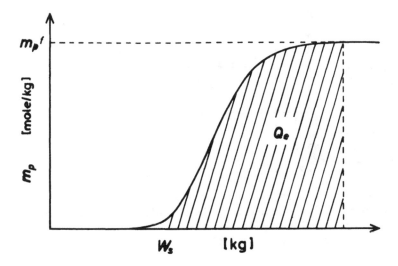

Fig. 3.3 Elution profile in frontal analysis.

Suzuki, 1974; Nakanishi et al., 1977b). An appropriate amount
of swollen ion exchanger is packed in a column (e.g., 1.5 cm
in diameter × 30 cm in length), and the column is equilibrated
by feeding an elution buffer at prescribed pH and ionic
strength. A small amount of the protein solution is then charged
onto the top of the column and eluted with the elution buffer.
After an appropriate time interval (retention time), a concentration profile, such as that shown in Fig. 3.4, is observed
at the exit of the column. The K value is determined by applying the moment method to the analysis of the elution curve
(Mehta et al., 1973; Suzuki, 1974; Nakanishi et al., 1977b).

The solution in the Laplace domain $\tilde{C}(Z, p)$ of the mass
balance model I shown in Sec. 2.1.3 is derived for pulse input
of the sample as

$$\tilde{C}(Z, p) = C_0 X_0 \tau \exp[f(Z, P)] \tag{3.42}$$

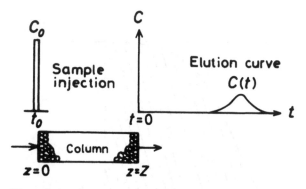

Fig. 3.4 General diagram of a pulse response experiment. A
sample solution with concentration C_0 is added during a period
t_0 to the top of a column. At the exit of the column, a concentration profile represented by $C(t)$ is observed.

where

$$f(Z, p) = (\delta/2) - [(\delta/2)^2 + \delta p \tau + (3HK\delta/\phi^2)g(Z, p)]^{1/2}$$

$$(3.43)$$

$$g(Z, p) = \phi(p\tau)^{1/2} \coth[\phi(p\tau)^{1/2}] - 1 \qquad (3.44)$$

$$\delta = \frac{Z}{D_L/u} \qquad \phi = \frac{R_p}{(\bar{D}_p\tau)^{1/2}} \qquad \tau = \frac{Z}{u}$$

$$X_0 = \frac{t_0}{\tau} \quad \text{and} \quad p = \alpha + \beta_i \qquad (3.45)$$

where p is a complex variable (for other notations see Chap. 2).

From Eqs. (3.42) through (3.45), the first normalized statistical moment μ'_1 (average retention time of a solution in a
column), for the pulse input is given by Eq. (3.46) using the
following relationship:

$$\mu'_n = \frac{(-1)^n \lim_{p \to 0} (d/dp)^n \tilde{C}(Z, p)}{\lim_{p \to 0} \tilde{C}(Z, p)}$$

$$\mu'_1 = \frac{Z}{u_0} [\varepsilon + (1 - \varepsilon)K] \qquad (3.46)$$

$$= \frac{Z}{u} (1 + HK) \qquad (3.46')$$

where $u_0 = u\varepsilon$ and $H = (1 - \varepsilon)/\varepsilon$.

When a high-molecular-weight substance that cannot enter the gel particle, such as Blue Dextran 2000 (molecular weight = 2×10^6) or Dextran T-2000 (molecular weight = 2×10^6), is used as a sample, Eq. (3.46) becomes

$$\mu'_1 = \left(\frac{Z}{u_0}\right) \varepsilon \qquad (3.47)$$

When an ion exchanger is positively charged, blue dextran cannot be used as a sample. The ε value is determined by measuring μ'_1 at various Z/u_0 values. Once ε is determined, the K value is obtained from Eq. (3.46). Figure 3.5 shows a plot of μ'_1 and Z/u_0 for agarose ion-exchange columns using ovalbumin or β-lactoglobulin solution as a sample.

3.3 EXPERIMENTAL DATA

3.3.1 Equilibria at Low Protein Concentrations

There are very few experimental data on ion-exchange equilibria or adsorption isotherms with proteins. In most cases the data were taken at low protein concentrations in the solution. Under these conditions the distribution coefficient K was found to be constant regardless of the protein concentration,

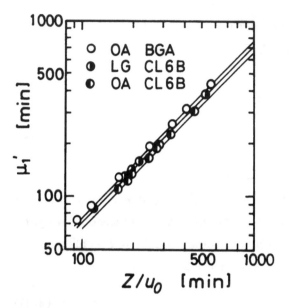

Fig. 3.5 Experimental relations between the first statistical moment and Z/u_0. OA, ovalbumin; LG, lactoglobulin; BGA, DEAE Bio-Gel A; CL6B, DEAE Sepharose CL6B. Column size, $1.5^\phi \times$ 30 cm. The K_{crt} values obtained from the results shown in the figure are summarized in Table 3.1. (From Yamamoto et al., 1983b.)

which is consistent with the models shown in the previous section [Eq. (3.10) or (3.33)]. Boardman and Partridge (1953, 1955) first studied the equilibrium relationship of such proteins as cytochrome c on ion-exchange resins (Amberlite IRC-50) by measuring the variation in the elution volumes with salt concentration. They explained elution behavior based on the law of mass action [Eq. (3.10)] and emphasized a strong dependency of the elution volume on the salt concentration, which becomes more significant as the electrochemical valence increases, as seen in Eq. (3.10). It was further pointed out that, in addition to electrostatic forces, the effect of short-range secondary forces

due to uncharged carboxylic groups cannot be ignored, as
shown by the following equations:

$$R^-Na^+ + H^+ \rightleftharpoons RH + Na^+ \qquad\qquad (3.48)$$

$$RH + P^+ \rightarrow RH\text{---}P^+ \qquad\qquad (3.49)$$

$$RH\text{---}P^+ + Na^+ \rightleftharpoons R^-Na^+ + P^+ + H^+ \qquad\qquad (3.50)$$

$$R^-Na^+ + P^+ \rightleftharpoons R^-P^+ + Na^+ \qquad\qquad (3.51)$$

Equation (3.48) shows the reversible exchange of Na^+ and H^+
ions; Eq. (3.49), the irreversible adsorption due to secondary
forces; Eq. (3.50), the desorption of the protein in the presence
of Na^+; and Eq. (3.51), the reversible adsorption of the protein
owing to electrostatic forces. Boardman and Partridge ascribed
the cause of the secondary forces to the concomitant removal
of a layer of electrostatically bound water. When only an electrostatic interaction exists, adsorption equilibria are principally
dependent on the net charge of the protein. When the secondary forces cannot be ignored, the equilibria are also influenced
by the chain length of the protein molecule, its configuration,
and the nature of the side chains other than the net charge.
Thus, it may be possible to separate proteins with the same net
charge but different molecular configurations by utilizing such
an effect, although a very strong interaction between the protein and resin causes denaturation.

Aranyi and Boross (1974) quantitatively analyzed the adsorption behavior of such proteins as ribonuclease A onto a weak
cation exchanger (Amberlite CG-50) based on the law of mass
action involving ion-exchange equilibria and hydrogen bonds.
They experimentally verified the linear relation between log K
and log (Na_s/Na_r) (where Na_s is the Na^+ concentration in the

solution and Na_r is the fixed ion concentration of the resin),
which is coincident with Eq. (3.10). They also showed a role
of the hydrogen bond in the adsorption of proteins onto the
ion exchangers. Kopaciewicz et al. [1983] have recently pointed
out the role of the charge asymmetry of the protein molecule in
the elution behavior of a column packed with ion-exchange
resins used for high-performance ion-exchange chromatography.

Adsorption isotherms of various proteins on hydrophilic ion
exchangers, such as DEAE-Sepharose CL6B, DEAE-Sephadex
A-25, DEAE-Sephadex A-50, DEAE-Bio-Gel A, and DEAE-Toyopearl 650, were measured by the authors (Yamamoto et al.,
1983b). Figure 3.6 shows the experimental results of the adsorption isotherms obtained either by a batch method or by
frontal analysis. As shown in Fig. 3.6, the tendencies of the
isotherms are dependent on the ionic strength, the pH, and the
types of ion exchangers and proteins. However, in all cases
the isotherms can be expressed as a linear relationship at sufficiently low protein concentrations. The effect of ionic strength
is shown for β-lactoglobulin A on DEAE-Sepharose CL6B. The
increase in the ionic strength from 0.11 to 0.16 brought about a
great change in the initial slope of the isotherm. With BSA on
DEAE-Sepharose CL6B, the decrease in pH from 7.9 to 6.1,
which means that the pH approached the pI (the pI of the BSA
is approximately 4.8), reduced the initial slope of the isotherms.
These results are qualitatively in fairly good agreement with
Eq. (3.10) or (3.33).

As the ionic strength increased, the initial quasi-linear part
of the isotherms became wider, and finally linear isotherms that
were independent of the ionic strength were obtained with
DEAE-Sepharose CL6B, DEAE-Bio-Gel, and DEAE-Toyopearl 650.

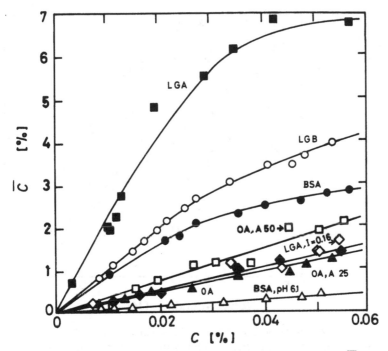

Fig. 3.6 Adsorption isotherms of proteins. Here, $\bar{C}$ and C denote the concentration of the protein in the ion exchanger and
the solution, respectively. Experimental conditions: pH 7.9,
I = 0.11 M, temperature 20°C, and the ion exchanger DEAE-
Sepharose CL6B unless otherwise indicated. The same abbreviations as in Table 3.1 are used for the sample proteins. Ion
exchangers: DEAE-Sephadex A-25, A 25; DEAE-Sephadex
A-50, A50.

Under these conditions, usually when the ionic strength of the
outer solution I > 0.5, electrostatic interaction between ion exchangers and proteins can be ignored and the ion exchanger can
be treated as gel chromatographic media, as mentioned previously. In order to investigate the dependency of the adsorption
equilibrium on ionic strength I, the distribution coefficients K
were calculated from the quasi-linear part of the isotherms over

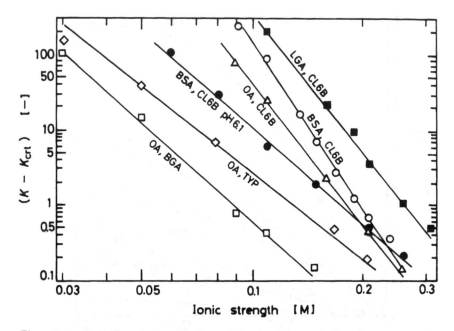

Fig. 3.7 Relation between K and ionic strength I. Here, pH is 7.9 unless otherwise indicated. The abbreviations for the sample proteins are the same as in Table 3.1. Ion exchangers: DEAE-Sepharose CL6B, CL6B; DEAE-Bio-Gel A, BGA; DEAE-Sephadex A-25, A 25; DEAE-Sephadex A-50, A 50; and DEAE-Toyopearl 650, TYP.

a wide range of ionic strength. As shown in Fig. 3.7, linear relations are found between the logarithms of $K - K_{crt}$ and that of I, where K_{crt} denotes the distribution coefficient obtained at the sufficiently high ionic strength by the pulse response experiment (Suzuki, 1974; Nakanishi et al., 1977b; Mehta et al., 1973). Thus, the K value is expressed as a function of I as

$$K = AI^B + K_{crt} \qquad\qquad (3.39)$$

The values of A, B, and K_{crt} are summarized in Table 3.1.

3.3.2 Equilibria at High Protein Concentrations

When the protein concentration in the outer solution increases, the adsorption isotherm usually exhibits a saturation curve. Figures 3.8 and 3.9 show adsorption isotherms of bovine serum albumin on two ion exchangers with different degrees of cross-linking, that is, Sephadex QAE A-25 and QAE A-50, at a constant pH 7.54. Sephadex QAE A-25 and QAE Q-50 are both strong anion exchangers with fixed ions of the quaternary aminoethyl group and with different degrees of cross-linking. Sephadex QAE A-25, originally prepared from Sephadex G-25, is more highly cross-linked than QAE A-50 (prepared from Sephadex G-50). As seen in Figs. 3.8 and 3.9, the adsorption isotherms for the two ion exchangers have different shapes. With Sephadex QAE A-50, the amount $\overline{C}$ of BSA adsorbed on the gel (g BSA per g dry resin) increases sharply with BSA concentration C in the solution at any ionic strength and seems saturated at relatively low concentrations. Furthermore, the $\overline{C}$ value in the

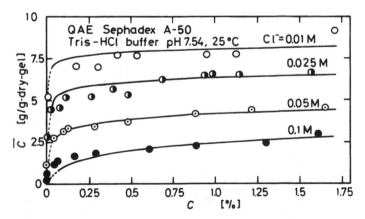

Fig. 3.8 Adsorption isotherms of BSA onto Sephadex QAE A-50 (calculated).

Table 3.1 Values of Parameters Appearing in Eq. (3.39) for Several Protein-Ion Exchanger Systems[a]

Ion exchanger[b]	d_{pa} (μm)[c]	Sample	K_{crt}	K'[d]	A	B	He
DEAE-Sepharose CL6B [145/200]	110 [93]	LGA	0.6	0.9	2.57×10^{-4}	-6.20	1.86
		LGB	0.6	0.9	9.41×10^{-5}	-6.28	
		OA	0.5	0.9	2.64×10^{-5}	-6.21	
		BSA	0.44	0.9	4.90×10^{-6}	-7.51	
	(pH 6.1)	BSA	0.44	0.9	6.12×10^{-4}	-4.33	
DEAE-Bio-Gel A [145/200]	123 [116]	OA	0.63	0.9	3.38×10^{-5}	-4.25	1.78
DEAE-Toyopearl 650 [200/250]	62 [83]	OA	0.52	0.87	4.75×10^{-4}	-3.76	1.56

DEAE-Sephadex A-25 [80/100]	151 [167]	OA	0	0.5^f	1.10×10^{-8}	-9.66	1.7
DEAE-Sephadex A-50	252	OA	0	1.0	1.83×10^{-3}	-4.49	2.57

[a]pH is 7.9 unless otherwise indicated and the following abbreviations are used: bovine serum albumin, BSA; ovalbumin, OA; β-lactoglobulin A, LGA; β-lactoglobulin B, LGB.

[b]The numeral in the bracket represents the sieve fraction, i.e., 145/200 means a fraction between 145 and 200 mesh size separated from original gels.

[c]The value of d_{pa} was measured for swollen gels at $I = 0.11$ and pH 7.9 and calculated according to Eq. (5.11). Values in brackets represent the d_p value for the original gel.

[d]The distribution coefficient K' for NaCl was determined by frontal analysis as mentioned in the text.

[e]$H = (V_t - V_0)/V_0$, where V_t and V_0 are the total and void volumes of the column, respectively.

[f]The value is determined by measuring the experimental results, where I_0 is 0.11 and I_{elu} is 0.31 at pH 7.9.

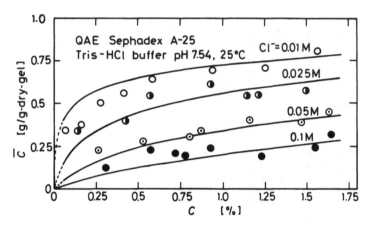

Fig. 3.9 Adsorption isotherms of BSA onto Sephadex QAE A-25 (calculated).

saturated state is strongly dependent on the ionic strength of the solution. On the other hand, with Sephadex QAE A-25, the BSA adsorbed on the gel ($\overline{C}$) gradually increases with the BSA concentration in the solution and this tendency is dependent on the ionic strength. Thus, it seems difficult to explain the experimental adsorption isotherms shown in Figs. 3.8 and 3.9 by the law of mass action. In order to quantitatively treat these adsorption isotherms an approach was made on the basis of the Donnan potential theory using data on the physical properties of the gel and protein, such as the concentration of the fixed ionic groups, swelling pressure, and electrochemical valence of the protein.

The concentration of the fixed ionic groups of the Sephadex QAE A-25 and 50 was determined as about 2.2 meq/g dry gel by titrating them with 0.5 N HCl at 25°C (the gels were dissolved in 1 M NaCl solution).

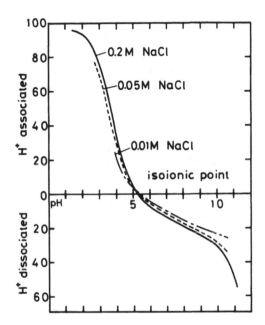

Fig. 3.10 Titration curves of BSA at 25°C for various NaCl concentrations. The dissociated or associated H$^+$ was calculated on the basis of the molecular weight of BSA = 65,000.

The relation between the electrochemical valence of BSA and pH was determined by titrating BSA in NaCl solution of various concentrations with HCl solution. The titration curves for BSA are shown in Fig. 3.10. Figure 3.10 shows that the isoionic point of BSA is around 5.2 and its negative charge increases with pH, depending slightly on NaCl concentration. Thus, the electrochemical valence of BSA can be calculated as shown in Fig. 3.11, where the change of charge due to the adsorption of small ions on the BSA molecules is ignored.

The swelling pressure of the ion exchanger was determined by measuring its volume change with the ionic strength of the NaCl solution according to the method proposed by Gregor (1951). The electroneutrality condition in the gel is

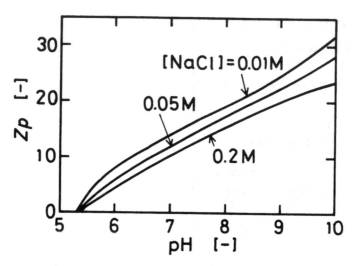

Fig. 3.11   Molecular charge of BSA estimated from Fig. 3.10.

$$\bar{n}_{Na} = \bar{n}_{Cl} - n_R \tag{3.52}$$

where $\bar{n}_{Na}$, $\bar{n}_{Cl}$, and n_R are moles of Na^+, Cl^-, and fixed cationic groups (QAE groups) in the gel, respectively. The sum of the movable ions in the gel is given as

$$\bar{n}_{Na} + \bar{n}_{Cl} = 2\bar{n}_{Cl} - n_R \tag{3.53}$$

Thus, the gel inside volume V_i becomes

$$V_i = \bar{n}_w v_w + v(2\bar{n}_{Cl} - n_R) \tag{3.54}$$

or

$$\frac{V_i}{n_R} = \left(\frac{\bar{n}_w}{n_R}\right) v_w + v\left(\frac{2\bar{n}_{Cl}}{n_R} - 1\right) \tag{3.54'}$$

Here, the relation of $v_{Na} = v_{Cl} = v$ (= 0.039 L/mol) is assumed for simplicity, and v_w (partial molar volume of water) is specified as 0.018 L/mol. The Gibbs-Donnan equation for the ideal case in which all activity coefficients are unity is written

$$\mathcal{R}T \ln \frac{\overline{X}_{Na}\overline{X}_{Cl}}{X_{Na}X_{Cl}} = -2\Pi v \qquad (3.55)$$

or

$$\mathcal{R}T \ln \frac{\overline{X}_{Na}\overline{X}_{Cl}}{X_{NaCl}^2} = -2\Pi v \qquad (3.55')$$

where $\overline{X}$ and X are the mole fractions in the gel and in the outer solution, respectively. Similarly, the thermodynamic os- motic pressure is calculated from the equation

$$\Pi = \frac{\mathcal{R}T}{v_w} \ln \frac{X_w}{\overline{X}_w} \qquad (3.56)$$

Furthermore, the dependency of the inner volume per mole of the fixed ionic groups V_i/n_R on Π may be expressed by Eq. (3.57), as postulated by Gregor (1948):

$$\frac{V_i}{n_R} = a\Pi + b \qquad (3.57)$$

The constant a reflects the elastic properties of the gel and should be smaller for more highly cross-linked gels. The values of a (L/atm mol^{-1}) and b (L/mol) in Eq. (3.57) can be determined to achieve the best fit between the experimental data of inner gel volume V_i and those calculated by Eqs. (3.54) through

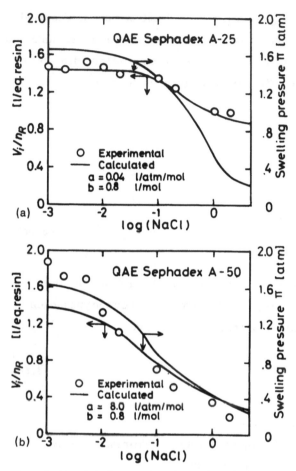

Fig. 3.12 (a) Dependencies of V_i/n_R and Π on NaCl concentration for Sephadex QAE A-25. (b) Dependencies of V_i/n_R and Π on NaCl concentration for Sephadex QAE A-50.

(3.57) for the NaCl concentration measured in the solution. As shown in Fig. 3.12a and b, the experimental results are in good agreement with the calculated results with a and b values, particularly for Sephadex QAE A-25. For Sephadex QAE A-50, experimental errors due to extraordinarily high swelling at low

ionic concentrations must be accounted for. The lower value of
a for the Sephadex QAE A-25 than that for the QAE A-50 indicates that the former has a more rigid gel structure than the
latter. The swelling pressure falls between 0 and 17 atm with
QAE A-25 and 0 and 1.6 atm with QAE A-50, respectively, depending on the NaCl concentrations in the solution. Thus, in
ion-exchange equilibria with the protein, the Π value can be
predicted by Eq. (3.57) when V_i/n_R is known. The experimentally obtained V_i/n_R value and the corresponding Π value for
Sephadex QAE A-25 as well as for QAE A-50 at an equilibrium
protein concentration of about 1% are shown in Table 3.2. The
decrease in V_i/n_R compared with that shown in Fig. 3.12a or
b is mainly ascribable to the decrease in repulsive forces inside
the ion exchanger due to large-scale adsorption of protein.
With the values of the parameters already determined, Eqs.
(3.23) through (3.32) were solved for the protein concentration
in the gel as a function of the Cl⁻ concentration and the protein concentration in the ionic solution. The suffixes P, A, and
B in Eqs. (3.23) through (3.32) correspond in this experiment
to BSA, Tris ion, and chloride ion, respectively. In order to
express the effect of the activity coefficient on the equilibria,
we took into consideration such interactions as ion-pair formation
between protein or Cl⁻ ion and the fixed ionic groups of the
ion exchanger. For simplicity, let us assume that these interactions are expressed by the equations

$$\overline{m}_{RP} = K_P \overline{m}_P m_{R^+} \tag{3.58}$$

$$\overline{m}_{RCl} = K_{Cl} \overline{m}_{Cl} m_{R^+} \tag{3.59}$$

Table 3.2 Experimental Values of $\bar{C}$, V_i/n_R, and Π for QAE-Sephadex A-25 and A-50[a]

Cl⁻ (M)	Sephadex QAE A-25				Sephadex QAE A-50			
	C (%)	$\bar{C}$ (g/g dry gel)	V_i/n_R (L/eq. resin)	Π (atm)[b]	C (%)	$\bar{C}$ (g/g dry gel)	V_i/n_R (L/eq. resin)	Π (atm)[b]
0.01	1.61	0.82	1.73	22.5	1.06	7.8	9.86	1.13
0.025	1.18	0.55	1.73	22.5	0.65	6.2	7.68	0.86
0.05	1.18	0.42	1.5	17.5	0.71	4.0	6.82	0.75
0.10	1.46	0.2	1.59	20	0.65	2.1	7.23	0.80

[a]Measured in Tris-HCl buffer, pH 7.54, at 25°C.
[b]Estimated from Figs. 3.12a and 12b.

where   RP, RCl = complex of protein or Cl⁻ and fixed ionic
                   groups, respectively

   K_P, K_{Cl} = association constants

   m_{R^+} = concentration of fixed ionic groups that do not par-
              ticipate in ion-pair formation

Thus, the mass balance equation for the fixed ionic groups is
expressed as

$$m_R = \overline{m}_{RP}|Z_P| + \overline{m}_{RCl} + m_{R^+} \qquad (3.60)$$

For simplicity we assume further that the electrical potential
in the outer solution phase is zero; that is, the Donnan potential is defined on the basis of the outer solution phase. Therefore, the electrochemical valence of the protein, corresponding
to the gel phase, can be used for the calculation.

As seen in Figs. 3.8 and 3.9, the calculated results are in
good agreement with the experimental results, although several
assumptions were made. In the calculation, the experimental
values measured for the specified condition (shown in Table 3.2).
are used for all the protein concentrations in the solution, at
the fixed ionic concentration. In Table 3.3, the values of
parameters used for the calculation are summarized. In Table
3.4 and 3.5 are summarized the concentrations of BSA, Cl⁻ ion
and Tris ion, and pH in the ion exchanger and in the outer
solution at an equilibrium BSA concentration in the outer solution
of 1%. With Sephadex QAE A-25, the difference in Cl⁻ and Tris ion
concentrations between the two phases is extraordinarily large,
and the pH in the gel phase is higher by about 1 unit than that
in the outer solution. On the other hand, with Sephadex QAE
A-50, these differences are much less than with QAE A-25. The
protein is more concentrated inside the gel, however, with QAE

Table 3.3 Values of Parameters Used for the Calculation of Adsorption Isotherms[a]

(Cl$^-$) (M)	Sephadex QAE A-25			Sephadex QAE A-50		
	K_P (M^{-1})	K_{Cl} (M^{-1})	$|Z_P|$	K_P (M^{-1})	K_{Cl} (M^{-1})	$|Z_P|$
0.01	0	26	20	0	11	13
0.025	0	3.9	20	0	6	14
0.05	0	2.5	19	0	4.8	15
0.10	0	0.07	19	0	1.0	15

[a]$v_P = 46.54$ L/mol; $v_H + v_{Cl} = 0.036$ L/mol; $v_T + v_{Cl} = 0.156$ L/mol; $v_H + v_{OH} = 0.0181$ L/mol.

Table 3.4 Some Parameters Estimated for Sephadex QAE A-25[a]

(Cl$^-$) (M)	($\overline{Cl^-}$) (M)[b]	(T$^+$) (M)	($\overline{T^+}$) (M)	($\overline{BSA}$) (%)[c]	pH
0.01	0.102	0.0063	0.00054	19.1	8.55
0.025	0.25	0.0156	0.00135	13.9	8.45
0.05	0.335	0.0313	0.00418	10.4	8.37
0.1	0.629	0.0625	0.00874	5.8	6.33

[a]The pH and BSA concentrations in the outer-phase solution were 7.54 and 1%, respectively.
[b]Free form in the ion exchanger.
[c]Based on the swollen gel.

Table 3.5   Some Parameters Estimated for Sephadex QAE A-50[a]

(Cl^-) (M)	$(\overline{Cl^-})$ $(M)^{b}$	(T^+) (M)	$(\overline{T^+})$ (M)	$(\overline{BSA})$ $(\%)^{c}$	$\overline{pH}$
0.01	0.0156	0.0063	0.0040	37.3	8.09
0.025	0.0365	0.0156	0.0113	37.6	7.77
0.05	0.0688	0.0313	0.0214	28.0	7.68
0.1	0.133	0.0625	0.0467	15.7	7.66

[a]The pH and BSA concentrations in the outer-phase solution
were 7.54 and 1%, respectively.
[b]Free form in the ion exchanger.
[c]Based on the swollen gel.

A-50 gel than with QAE A-25 gel owing to the lower swelling
pressure.

   We have shown one approach for the quantitative treatment of
ion-exchange equilibria with proteins.   Although the experimental
data may be fit well by the simple treatment adopted here, this
is only one aspect of the equilibrium relation of protein ion exchange.   There are few available data on the interaction between
proteins and the fixed ionic groups of the ion exchangers.   In
the treatment here, the change in Γ_i between the ion exchanger
and the outer solution was ignored, and the effect of Γ_i on
equilibria was represented only by the interaction between proteins or chloride ion and fixed ionic groups.   It is also ques-
tionable whether the swelling pressure influences the ion-exchange behavior to a great extent, as shown earlier, since there
are no data on swelling pressure except ours.   Information on
the activity coefficients is also lacking, particularly for proteins.

   For engineering purposes, one of the empirical equations
shown in Sec. 3.1.3 may be used.   Equation (3.39) may be

particularly useful when low protein concentrations and shrink-
age of the gel at higher ionic strengths can be ignored. Over
a limited range of concentrations, the experimental adsorption
isotherms of proteins on ion exchange columns may be fitted by
a Langmuir equation (Yamamoto et al., 1983b; Tsou and Graham,
1985; Chase, 1985).

4
Diffusion in the Ion Exchanger

As described in Chap. 2, diffusion in the stationary phase, that is, intraparticle diffusion in the ion exchanger, is also responsible for dispersion of a solute in a chromatographic column. The residence time of a solute in the ion exchanger is distributed around an average value of $t_D = R_p^2/2\overline{D}$ (Giddings, 1965), where R_p is the radius of the ion exchanger and $\overline{D}$, the diffusion coefficient in the ion exchanger. The degree of the dispersion in a column becomes larger with an increase in the ratio of the residence time of a solute in the ion exchanger to the time required for passing through the interstices of a column. In this chapter, we provide information about diffusion in ion exchangers and a prediction method.

4.1 THEORETICAL TREATMENT

Ion exchangers are porous bodies composed of matrices with fixed ionic groups as a solid phase and pore interstices. Diffusion in the ion exchanger is thus greatly influenced by the structure of three-dimensional matrices in a way similar to that

in other porous materials. However, the diffusion of ionic substances in the ion exchanger is in principle much more complicated than that of nonionic substances. Its complexity mainly
arises because at least two ionic species participate in the ion-
exchange reaction and their behavior is interconnected, maintain-
ing electroneutrality at any position inside the ion exchanger.
Since such a process cannot be described by a single constant
diffusion coefficient, its mathematical treatment becomes difficult.
Differences in ionic concentration, pH, and swelling pressure in
the ion exchanger and the outer solution (see Chap. 3) also
make the problem complex. However, the ion-exchange chromatography of protein is characterized, as discussed in Chap. 1,
by the following: (1) the diffusion of salt ions is much faster
than that of proteins (by 10−100 times), and (2) the protein
concentration in the gel is considerably lower than that in the
salt. With these two characteristics of ion-exchange chromatography, the problem becomes simplified. Since the diffusional
velocity of the salt is very high, one may assume that an equilibrium state always exists in the ion exchanger with respect to
the salt concentration. Second, the effect of electrical transference caused by the electrical potential gradient can be ignored
because of lower protein concentrations in the ion exchanger, as
shown in this chapter.

Generally, the flux J_i of the ith species in the ion exchanger
may be given by the sum of the pure (statistical) diffusion due
to the concentration gradient $(J_i)_{diff}$ and electrical transference
$(J_i)_{el}$ caused by the electrical potential gradient as (Helfferich,
1962)

$$J_i = (J_i)_{diff} + (J_i)_{el} \tag{4.1}$$

$(J_i)_{diff}$ and $(J_i)_{el}$ are expressed by Eqs. (4.2) and (4.3), respectively, as

$$(J_i)_{diff} = -\overline{D}_i \text{ grad } \overline{C}_i \qquad (4.2)$$

$$(J_i)_{el} = -\frac{\overline{D}_i \mathfrak{F}}{\mathfrak{R}T} Z_i \overline{C}_i \text{ grad } \overline{\phi} \qquad (4.3)$$

where $\overline{D}_i$ and $\overline{C}_i$ are the diffusion coefficient in the ion exchanger and the concentration of the ith component based on the whole volume of the ion exchanger. From Eqs. (4.1) through (4.3) we obtain a so-called Nernst-Planck equation (1888, 1889, and 1890) in the ion exchanger as

$$J_i = -\overline{D}_i \left(\text{grad } \overline{C}_i + Z_i \overline{C}_i \frac{\mathfrak{F}}{\mathfrak{R}T} \text{ grad } \overline{\phi} \right) \qquad (4.4)$$

Here it is assumed for simplicity that the effects of activity coefficients, pH gradient, and pressure gradient are ignored. Equation (4.4) can be solved under two restrictions in the ion exchanger, that is, the electroneutrality condition [Eq. (4.5)] and the nonelectrical current condition [Eq. (4.6)], as

$$\Sigma \omega_i |Z_i| \overline{C}_i = -\omega C_R = \text{constant} \qquad (4.5)$$

$$\Sigma \omega_i |Z_i| J_i = 0 \qquad (4.6)$$

where ω_i and ω indicate signs of ionic species and ionic species of fixed ionic groups: -1 for anion and $+1$ for cation. In the simplest case with two counterions (a protein P and a counterion of a salt B) in the absence of coions,* the protein flux J_P in the ion exchanger becomes, from Eqs. (4.4) through (4.6) (Helfferich, 1962),

---

*For an ion-exchange equilibrium of a strong electrolyte A^+B^-,

$$J_P = - \frac{\overline{D}_P \overline{D}_B (Z_P^2 \overline{C}_P + Z_B^2 \overline{C}_B)}{Z_P^2 \overline{C}_P \overline{D}_P + Z_B^2 \overline{C}_B \overline{D}_B} \text{ grad } \overline{C}_P \qquad (4.7)$$

Thus the diffusion coefficient of protein (interdiffusion coefficient) is expressed as

$$\overline{D}_{PB} = \frac{\overline{D}_P \overline{D}_B (Z_P^2 \overline{C}_P + Z_B^2 \overline{C}_B)}{Z_P^2 \overline{C}_P \overline{D}_P + Z_B^2 \overline{C}_B \overline{D}_B} \qquad (4.8)$$

$\overline{D}_{PB}$ depends not only on the concentrations of the two species but also on their self diffusion coefficients and electrochemical valences.

When the protein concentration $\overline{C}_P$ in the ion exchanger is much lower than $\overline{C}_B$ found in the usual ion-exchange chromatography, Eq. (4.7) becomes

$$J_P = - \overline{D}_P \text{ grad } \overline{C}_P \qquad (4.9)$$

Equation (4.9) states that the flux of a protein inside the ion exchanger is simply given by a product of the intraparticle diffusion coefficient of the protein and its concentration gradient. In other words, the diffusion of proteins in the ion exchanger

─────────────

(continued from page 147)
the following equation is derived for the distribution coefficient of the coion (Helfferich, 1962) from the Donnan potential theory:

$$K_B = \frac{\overline{m}_B}{m_B} = \left[\left(\frac{m_R}{2Z_B m_B} \right)^2 + \left(\frac{\gamma}{\overline{\gamma}} \right)^2 \left(\frac{\overline{a}_w}{a_w} \right)^{\overline{v}_{AB}/v_w} \right]^{1/2} - \frac{m_R}{2|Z_B| m_B} \qquad (4.10)$$

From Eq. (4.10), it is easily found that K_B approaches unity as the m_B value increases.

can be treated as simple diffusion into the porous matrices
equilibrated at a specified salt concentration, without taking
into consideration the effect of electrical charge.

The $\overline{D}_P$ value in Eq. (4.9) depends on various factors, such
as the ionic strength of the solution, the degree of cross-linking
of the ion exchanger, and the size of the protein. We now examine the content of $\overline{D}_P$ in more detail. As described previously,
an ion exchanger consists of matrices with fixed ionic groups
(solid phase) and interstices among them (pores). It may be
assumed that protein molecules in the ion exchanger are partly
adsorbed to the surface of the solid phase and exists partly in
the pore liquid. Thus, the protein concentration on the basis
of a whole gel volume $\overline{C}_P$ is given by

$$\overline{C}_P = (1 - \varepsilon_p)\overline{C}_P{}^q + \varepsilon_p\overline{C}_P{}^P \qquad (4.11)$$

where $\overline{C}_P{}^q$ and $\overline{C}_P{}^P$ are the protein concentration in the solid
phase and pore liquid, respectively, and ε_p is the void fraction
of the pore. $\overline{C}_P{}^P$ may be expressed by a linear relation with
the concentration at the gel particle surface in the outer solution C_P in the region of low protein concentrations (see Chap.
3) as

$$\overline{C}_P{}^P = K_{ps}C_P \qquad (4.12)$$

Furthermore, it is assumed that the equilibrium relation between
$\overline{C}_P{}^q$ and $\overline{C}_P{}^P$ is given as

$$\overline{C}_P{}^q = K_{pq}\overline{C}_P{}^P \qquad (4.13)$$

From Eqs. (4.11) and (4.13), $\overline{\overline{C}}_P$ is expressed by

$$\overline{C}_P = [(1 - \epsilon_p)K_{pq} + \epsilon_p]\overline{C}_P^{\,P}$$

$$= [(1 - \epsilon_p)K_{pq} + \epsilon_p]\, K_{ps}C_P \tag{4.14}$$

On the other hand, the experimentally obtained distribution coefficient K should also be constant regardless of the protein concentration:

$$\overline{C}_P = KC_P \tag{4.15}$$

Accordingly one obtains, from Eqs. (4.14) and (4.15),

$$K = K_{ps}[(1 - \epsilon_p)K_{pq} + \epsilon_p] \tag{4.16}$$

The diffusion flux of the protein per unit area of the ion exchanger is expressed as follows when the surface diffusion of the protein on the solid phase is ignored:

$$J_P = -\overline{D}_P \frac{\partial \overline{C}_P}{\partial r} = -\overline{D}_P^{\,P} \epsilon_P \frac{\partial \overline{C}_P^{\,P}}{\partial r} \tag{4.17}$$

where $\overline{D}_P^{\,P}$ is the diffusion coefficient of the protein with respect to the pore liquid. By rearranging the right-hand side, Eq. (4.17) becomes, using Eqs. (4.12) and (4.15),

$$J_P = -\overline{D}_P^{\,P} \epsilon_p \frac{K_{ps}}{K} \frac{\partial \overline{C}_P}{\partial r} \tag{4.18}$$

Thus,

$$\overline{D}_P K = \overline{D}_P^{\,P} \epsilon_p K_{ps} \tag{4.19}$$

In Eq. (4.19), the diffusion coefficient $\overline{D}_P^{\,P}$ in the pore liquid is expressed by $\overline{D}_P^{\,P} = D_m \times f$ (structure of matrix), where D_m is the molecular diffusion coefficient and K_{ps} is a function of the

Donnan potential and the structure of the matrix. When the concentration difference in the ion exchanger and outer solution caused by the Donnan potential is negligible, all the terms on the right-hand side of Eq. (4.19) are simply dependent on the physical structure of the gel and protein. Thus,

$$\bar{D}_p K = \text{constant} \quad \text{for a fixed ion exchanger and protein} \quad (4.20)$$

Equation (4.20) can be used to predict $\bar{D}_p$ at any ionic strength using the value determined at a fixed ionic strength when the ion exchanger does not deform with ionic strength. Such conditions may be realized with the ion exchanger having a low fixed ion concentration and degree of cross-linking (small swelling pressure). It should be noted here that diffusion in the ion exchanger is closely related to the mechanism of ion-exchange equilibrium.

4.2 EXPERIMENTAL METHODS

Diffusion coefficients of a protein in the ion exchanger can be easily determined either by a batch method or by a pulse response method using a column, with the assumption of a constant distribution coefficient.

4.2.1 A Batch Method

In this method, the diffusion coefficient is determined by analyzing the non-steady state of the adsorption or desorption process of the protein into or from the ion exchanger, respectively. Because of the simplicity of the experimental procedure, it is preferable to follow the adsorption process. To an appropriate amount of the ion exchanger W, equilibrated with the assay buf-

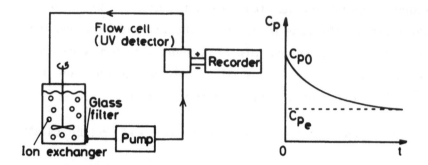

Fig. 4.1 Apparatus for measurement of the adsorption of protein to the ion exchanger.

fer, a protein solution of known concentration is added. The solution is mixed with a magnetically rotated bar. The decrease in the protein concentration in the outer solution due to the adsorption is analyzed either by pipetting through a glass filter or continuous monitoring by ultraviolet (UV) flow cell (Fig. 4.1).

As shown in Fig. 4.1, the protein concentration in the outer solution C_p gradually decreases as the time elapsed until the equilibrium state. Such an adsorption process is analyzed as follows. When the mixing of the solution is sufficiently strong, mass transfer resistance through a thin stagnant film adjacent to the ion exchanger can be ignored. The mass balance equation of the protein in the spherical ion exchanger with radius R_p is given by Eq. (4.21):

$$\frac{\partial \overline{C}_P}{\partial t} = \overline{D}_P \left(\frac{\partial^2 \overline{C}_P}{\partial r^2} + \frac{2}{r} \frac{\partial \overline{C}_P}{\partial r} \right) \tag{4.21}$$

Equation (4.21) is solved under the following initial and boundary conditions

$$\overline{C}_P = 0 \qquad R_p \geq r \geq 0, \; t = 0 \tag{4.22}$$

$$C_P = C_{P0} \qquad r > R_p, \; t = 0 \tag{4.23}$$

$$\frac{\partial \overline{C}_P}{\partial r} = 0 \qquad r = 0, \; t > 0 \tag{4.24}$$

$$\left(V_t - \frac{W}{\rho_g}\right)\frac{dC_P}{dt} = -\frac{3W}{R_p \rho_g}\,\overline{D}_P\left(\frac{\partial C_P}{\partial r}\right) \quad r = R_p, \; t > 0 \tag{4.25}$$

$$\overline{C}_P = KC_P \tag{4.15}$$

where V_t = total volume of a protein solution (solution plus
             ion exchanger)

W = weight of an ion exchanger

C_{P0} = initial protein concentration in the solution

ρ_g = density of the ion exchanger (approximately 1.0)

Equations (4.21) through (4.25) with Eq. (4.15) are solved as
(Crank, 1975)

$$M = \frac{C_{P0} - C_P}{C_{P0} - C_P} \tag{4.26}$$

$$M = 1 - \sum_{k=1}^{\infty} \frac{\exp(-\lambda_k^2 \overline{D}_P t / R_p^2)}{(3\alpha/2) + \lambda_k^2/6(1 + \alpha)} \tag{4.27}$$

where

$$\alpha = \frac{KW}{(V_t - \overline{W}/\rho_g)\rho_g} \tag{4.28}$$

and λ_k is the kth root of λ in Eq. (4.29):

$$\lambda^2 = 3\alpha \left(\frac{\lambda}{\tan \lambda} - 1 \right) \tag{4.29}$$

The equilibrium protein concentration C_{Pe} in the solution and the distribution coefficeint K are related as follows:

$$KC_{Pe} = \left(\frac{(C_{P0} - C_{Pe})(V_t - \overline{W}/\rho_g)}{W/\rho_g} \right) \tag{4.30}$$

When $(\overline{D}_p t / R_p^2)^{1/2}$ is smaller than 0.1, Eq. (4.27) is approximated to the following equation (Carman and Haul, 1954):

$$
\begin{aligned}
M &= \frac{C_{P0} - C_P}{C_{P0} - C_{Pe}} \\
&= \left(1 + \frac{1}{\alpha} \right) \left(1 - \frac{\gamma_1}{\gamma_1 + \gamma_2} \text{ e erfc } [3\alpha\gamma_1 (\overline{D}_p t / R_p^2)^{1/2}] \right. \\
&\quad \left. - \frac{\gamma_2}{\gamma_1 + \gamma_2} \text{ e erfc } [-3\alpha\gamma_2 (\overline{D}_p t / R_p^2)^{1/2}] \right)
\end{aligned}
\tag{4.31}
$$

where

$$\gamma_1 = \frac{1}{2} \left[\left(1 + \frac{4}{3\alpha} \right)^{1/2} + 1 \right] \tag{4.32}$$

$$\gamma_2 = \gamma_1 - 1 \tag{4.33}$$

$$\text{e erfc}(x) = \exp(x^2) \text{ erfc}(x) \tag{4.34}$$

Equation (4.27) is further simplified if the volume of the solution is so large that C_p = constant (= 0). This condition is more easily realized in the experiment by following the desorption process of protein from the ion exchanger than by following the adsorption process. In this case, all of the protein is in the ion exchanger, with a uniform concentration initially and no protein in the solution:

$$\overline{C}_P = \overline{C}_{P0} \qquad R_p \geqslant r \geqslant 0, \ t = 0 \qquad\qquad (4.35)$$

$$C_P = 0 \qquad r > R_p, \ t = 0 \qquad\qquad (4.36)$$

Equation (4.36) is also valid when $t > 0$. Solution of Eq. (4.21) under the conditions just described is

$$M = 1 - \frac{\overline{C}_P}{\overline{C}_{P0}} = 1 - \frac{6}{\pi^2} \sum_{k=1}^{\infty} \frac{1}{k^2} e^{-\overline{D}_p t \pi^2 k^2 / R_p^2} \qquad\qquad (4.37)$$

The $\overline{D}_p$ value can be determined to obtain the best fit between the experimental M values and the values calculated by Eq. (4.27), (4.31), or (4.37), depending on the experimental conditions. Although the batch method described here is very simple, it requires rapid and accurate measurement of the protein concentration in the solution. The salt concentration at which protein is eluted in IEC is as large as 0.1–0.5 M, as shown in Chap. 8, and the corresponding K values are less than 5.0. Under such conditions the $\overline{D}_p/R_p^2$ value is relatively large and as a result an equilibrium state is reached in very short time, as seen in Eqs. (4.27), (4.31), and (4.37). In this case, a pulse response method (see also Chap. 3) is useful

4.2.2 Pulse Response Method

The pulse response experiment also gives the $\overline{D}_p$ value, particularly when the K value is low in the presence of a high ionic strength of the buffer solution. The second central moment μ_2 (the variance of the peak) for the pulse input is derived in a way similar to the derivation of μ_1 (Chap. 4):

$$\mu_2 = 2 \frac{Z}{u} \left[\frac{D_L}{u^2} (1 + HK)^2 + \frac{HKR_p^2}{15\overline{D}_p} \right] \qquad\qquad (4.38)$$

where $u_0 = u\varepsilon$ and $H = (1 - \varepsilon)/\varepsilon$. Equation (4.38) is re-arranged as

$$\frac{\mu_2}{2Z/u} = \left[\frac{D_L}{u}\,(1 + HK)^2\,\frac{1}{u} + \frac{HKR_p^{\,2}}{15\overline{D}_P}\right] \tag{4.39}$$

When D_L/u is constant, as shown later (Chap. 5), plotting $\mu_2/(2Z/u)$ against $1/u$ gives a straight line, and $\overline{D}_P$ is calculated from the intercept of the line as

$$\overline{D}_P = \frac{HR_p^{\,2}K}{15 \times \text{intercept}} \tag{4.40}$$

It should be noted that the $\overline{D}_P$ value determined by this method includes a large error when the elution curve is asymmetric because the tailing part of the elution curve contributes considerably to the μ_2 value (Yamamoto et al., 1978). We have shown by computing the elution curves for various combinations of the values of parameters that symmetrical peaks are obtained when $\phi = R_p/(\overline{D}_P Z/u)^{1/2} < 0.2$ (see Sec. 2.1.6) (Yamamoto et al., 1979). Furthermore, $\overline{D}_P$ values tend to be lower than those obtained by the batch method, possibly because of a so-called channeling effect (Nakanishi et al., 1977b; Suzuki and Smith, 1972; Langer et al., 1978).

4.3 EXPERIMENTAL DATA

Only a few data are reported on the diffusion coefficient of proteins in the ion exchanger. Adachi et al. (1978) measured the $\overline{D}_P$ of glucoamylase in cation exchangers SP-Sephadex C-25 and C-50 by the batch method at a low protein concentration region where a linear relation of adsorption isotherm holds. In

Table 4.1 Distribution and Diffusion Coefficients of Glucoamylase in Ion Exchanger SP-Sephadex C-50 at 25°C and pH 5.0

d_p (μm)	Acetate buffer (M)	K	$\overline{D}_P$ (cm^2/s)	$K\,\overline{D}_P$ (cm^2/s)
220	0.01	3.0×10^4	7×10^{-11}	2×10^{-6}
190	0.05	2.5×10^2	2×10^{-9}	0.5×10^{-6}

Source: From Adachi et al. (1978).

Table 4.1, the $\overline{D}_P$ values determined for SP-Sephadex C-50 are listed. The $\overline{D}_P$ values are extraordinarily low in comparison with the molecular diffusion coefficient (approximately 1.3×10^{-6} cm^2/ s) and decreased with the increase in the K value. However, the products K and $\overline{D}_P$ take roughly a constant value [see Eq. (4.20)] and are in the order of the molecular diffusion coefficient in spite of a large difference in K values. These extremely low diffusion coefficients $\overline{D}_P$ cannot be explained simply by the steric hindrance of gel matrix but are largely ascribed to the binding of the enzyme to the fixed ionic groups; that is, the bound enzyme cannot diffuse on the surface of the solid phase—only the free enzyme can diffuse in the ion exchanger. Using the pulse response method, we measured the $\overline{D}_P$ of various proteins in several ion exchangers at such high ionic strengths I_{crt} that the electrostatic interaction is negligible between proteins and fixed ionic groups in the ion exchanger (Yamamoto et al., 198b). The $\overline{D}_{P,crt}$ values are shown in Table 4.2 with the distribution coefficient K_{crt} obtained under the same conditions. The $\overline{D}_{P,crt}$ values determined are lower than the molecular diffusion by a factor of 1/3 to 1/10. Since

Table 4.2 Values of $\bar{D}_{crt}$ and K_{crt} at 20°C and pH 7.9 for Various Ion-Exchange Columns

Ion exchanger[a]	d_{pa} (μm)[b]	Sample[c]	K_{crt}	$\bar{D}_{crt} \times 10^7$ (cm²/s)	$\bar{D}_{crt}/D_m$[d]
DEAE-Sepharose CL6B [145/200]	110	LGA	0.6	2.1	0.27
		LGB	0.6	2.1	0.27
		OA	0.5	1.8	0.24
		BSA	0.44	1.4	0.23
DEAE-Bio-Gel A [145/200]	123	OA	0.63	1.9	0.26
DEAE-Toyopearl 650 [200/250]	62	OA	0.52	1.4	0.19

[a]The numeral in the bracket represents the range of sieving, e.g., 145/200 means a fraction between 145 and 200 mesh sizes separated from original gels.
[b]The value of d_{pa} was measured for swollen gels at I = 0.11 and pH 7.9 and calculated according to Eq. (5.11).
[c]The following abbreviations are used for proteins: bovine serum albumin, BSA; ovalbumin, OA; β-lactoglobulin A, LGA; β-lactoglobulin B, LGB.
[d]The molecular diffusion coefficients D_m at 20°C are shown in Appendix F.
Source: Yamamoto et al. (1983b).

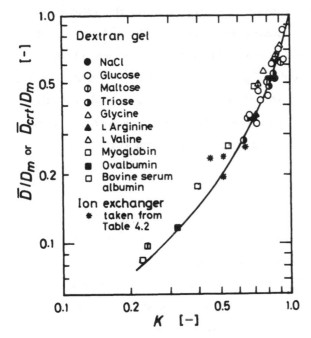

Fig. 4.2 Relationship between $\overline{D}/D_m$ and K for dextran gels and ion exchangers. The temperature is 25°C unless otherwise specified. For the values of D_m, see Appendix F. (From Nakanishi et al., 1983.)

in this case the binding of the protein to the fixed ionic groups is negligible, lowering of the diffusion coefficient in the ion exchanger is ascribable to the steric hindrance of the gel matrix. The effect of steric hindrance of the gel may be related to the distribution coefficient. We measured the diffusion coefficients of various substances in a single dextran gel particle by the single bead method (Horowitz and Fenichel, 1964; Nakanishi, 1977a), and a good correlation between $\overline{D}/D_m$ and K was obtained, as shown in Fig. 4.2, regardless of the types of diffusing substances. The $\overline{D_p}/D_m$ values shown in Table 4.2 are also plotted

in Fig. 4.2, and they appear to scatter around the solid curve,
although the data may include the effect of channeling, as pointed
out previously. Figure 4.2 may be useful for rough prediction
of the diffusion coefficeint $\overline{D}_{P,crt}$ in the gel at the critical condition. The $\overline{D}_P$ value at any ionic constant is obtained from the
following relation [see Eq. (4.20)] when the gel does not shrink
or swell owing to the change in the ionic strength:

$$\overline{D}_{P,crt}K_{crt} = \overline{D}_P K \qquad\qquad (4.41)$$

When K_{crt} is actually zero, as in the case of DEAE-Sephadex
A-25, the $\overline{D}_P$ value should be measured by the pulse response
experiment under such conditions that $K \cong 1$.

The diffusion coefficient at low protein concentrations has
been discussed. When the protein concentration becomes higher,
we should also take into consideration the interaction between
proteins themselves and between proteins and low-molecular-
weight ions, the effect of swelling pressure, and the change in
pH between the inner phase of the ion exchanger and the outer
solution. Under such conditions, the diffusion coefficients for
adsorption onto the ion exchanger and that for desorption from
it may be different. Furthermore, the gel must shrink or swell
more or less during adsorption or desorption. Graham and Fook
(1982) and Tsou and Graham (1985) examined the protein adsorption behavior on the ion exchanger at high protein concentrations
by using a simple two phase resistance model.

There are many problems yet to be examined.

5

Axial Dispersion in the Ion-Exchange Column

Axial dispersion of a solute in a column is also responsible for the dispersion of a solute in a chromatographic column as well as intraparticle diffusion (Chap. 4). In the usual case, the axial dispersion in an ion-exchange chromatographic column can be treated in a way similar to that in a gel chromatographic column or in a column packed with nonporous packing materials. According to the mass balance model I given in Sec. 2.1.3, the mass balance equation in a column with porous gels is given by Eq. (5.1) (the effect of mass transfer across the mobile/gel phase interface is ignored) with appropriate initial and boundary conditions:

$$\frac{\partial C}{\partial t} = D_L \frac{\partial^2 C}{\partial z^2} - u \frac{\partial C}{\partial z} - \frac{3H\overline{D}}{R_p} \left. \frac{\partial \overline{C}}{\partial r} \right|_{r=R_p} \tag{5.1}$$

Equation (5.1) is rearranged in nondimensional form as

$$\frac{\partial C^*}{\partial t^*} = \frac{1}{Pe} \frac{\partial^2 C^*}{\partial z^{*2}} - \frac{\partial C^*}{\partial z^*} - \frac{12H}{\overline{Pe}} \left. \frac{\partial \overline{C}^*}{\partial d^*} \right|_{d^*=1} \tag{5.2}$$

161

where $t^* = tu/d_p$

$C^* = C/C_{ref}$ (C_{ref} is any fixed concentration)

$Pe = d_p u/D_L$

$z^* = z/d_p$

$\overline{Pe} = d_p u/\overline{D}$

$d^* = d/d_p = 2r/d_p$

The Peclet number $Pe = d_p u/D_L$ indicates the degree of axial dispersion in a column. The higher the Pe number, the smaller is the dispersion of a solute in a column. For a column with nonporous packing materials, the third terms on the right-hand side of Eqs. (5.1) and (5.2) are eliminated. In this case the dispersion of a solute (second central moment μ_2) in a column is expressed as

$$\mu_2 = \frac{2Zd_p}{Peu^2} \tag{5.3}$$

where Z = length of a column. On the other hand, μ_2 is given by Eq. (5.4) according to the plate theory (see Secs. 2.1.5 and 2.1.6) (Martin and Synge, 1941).

$$\mu_{2,p} = \frac{(Z/u)^2}{N_p} \tag{5.4}$$

where N_p is the number of plates.

By equating Eqs. (5.3) and (5.4), one obtains the relationship

$$N_p = \frac{(Z/d_p)Pe}{2} \tag{5.5}$$

Equation (5.5) means that the Pe number is related to the plate theory by the length of the actual mixing times or number of a

plate in a column with respect to Z/d_p (the maximum number
of mixing times). Thus, previous data on axial dispersion in
a chromatographic column were in most cases represented as a
function of the Pe number.

Axial dispersion in a chromatographic column is mainly caused
by the two factors: (1) molecular diffusion along a column
length and (2) nonuniformity of linear velocity in the mobile
phase at the cross section of a column (eddy diffusion). Therefore, the axial (longitudinal) dispersion coefficient D_L is expressed by the sum of both factors as

$$D_L = \gamma_m D_m + \lambda u d_p \qquad (5.6)$$

or

$$\frac{1}{Pe} = \frac{\gamma_m}{\nu} + \lambda \qquad (5.7)$$

where

$$\nu = \frac{u d_p}{D_m} \qquad (5.8)$$

In Eq. (5.6), γ_m is a so-called labyrinth factor, the correction
factor for the tortuous flow channels, and λ is that for eddy
diffusion. ν is the nondimensional reduced velocity. Except
when flow velocity u is very low, the effect of the longitudinal
molecular diffusion can be neglected. When one microscopically
observes the phenomena taking place in a column, there are
several causes for the nonuniformity of the linear velocity.
Giddings (1965) pointed out five factors for the causes of this
velocity nonuniformity: (1) a transchannel effect due to a
velocity nonuniformity in each interstitial flow channel, (2) a
transparticle effect due to the existence of the stagnant mobile

phase, (3) a short-range interchannel effect caused by a difference in the packing state on a relatively small scale, (4) a
long-range interchannel effect, and (5) a transcolumn effect
caused by a velocity difference on a columnwide scale. Thus,
the Pe number can easily be expected to depend on many factors,
such as flow velocity, particle size, shape of the gel particle,
column diameter and length, the ratio of particle diameter to
column diameter, gel particle size distribution, void fraction of
a column, the packing method (by which the nonuniformity of a
packing state changes), and properties of elution buffer (viscosity, density, and so on). Moreover, the Pe number sometimes depends on the measuring method (Cluff and Hawkes,
1976). In order to assess the effect of velocity on the Pe number, the Reynolds number Re is often used in chemical engineering, particularly for the gas-solid system. However, in
column chromatography, including gel, ion-exchange, and affinity chromatography, the reduced velocity ν is preferably
used in place of the Re number. The Re number may be a
useful parameter in studying dispersion in a column as a function of turbulence, since it is a measure of the degree of turbulence (Cluff and Hawkes, 1976). Cluff and Hawkes (1976)
collected 750 published data obtained with a column packed with
nonporous spherical beads and recalculated to yield a plot of
$2/Pe$ versus ν, as shown in Fig. 5.1. They further analyzed
the scatter of the data in terms of the column parameters other
than ν, such as the Re number ($= ud_p\rho/\mu$), the column diameter,
and the particle size. As seen in Fig. 5.1, the $2/Pe$ value
takes a higher value at lower values of ν, particularly when
$\nu < 1$, where the effect of the molecular diffusion cannot be

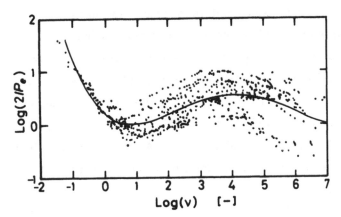

Fig. 5.1 Relationship between log (2/Pe) and log ν. [Reproduced from the Journal of Chromatographic Science by permission of Cluff and Hawkes (1976) and Preston Publications, Inc.]

ignored. Under the usual conditions of ion-exchange chromatography, ν is in the range 10–200, where 2/Pe seems to be independent of ν, although there is a large degree of data scattering. Furthermore, in this range most of the data take 0 to 0.3 as the values of log (2/Pe); that is, Pe = 1–2. It should be noted that the Pe number obtained from a column packed with small particles in a gas-solid system has a tendency different from that in the liquid-solid system shown here. In the gas-solid system, the D_L/u value becomes constant regardless of d_p, which is caused by channeling of the flow inside a column, when d_p is smaller than 1 mm (Suzuki and Smith, 1972). The Pe numbers obtained for chromatographic columns are appreciably lower than those for columns packed with larger particles even in a liquid-solid system. The lower Pe number in the chromatographic column can be ascribed to the much smaller

particle size of the gel used, and thus great caution is taken
in packing gels and applying the sample to the column.

We have also measured D_L for the column packed with several Sephadex gels (gel filtration chromatography media) or ion
exchangers by the pulse response method described in Chaps.
3 and 4. Under the experimental conditions (Re = 0.1−10),
D_L/u was found to be independent of u. As described in Chap.
4, the second central moment of the elution curve is related to
1/u by the equation

$$\frac{\mu_2}{2Z/u} = \frac{D_L}{u} \frac{(1 + HK)^2}{u} + \frac{HR_p^2 K}{15\overline{D}} \tag{5.9}$$

Plotting $\mu_2/(2Z/u)$ against 1/u gives a straight line, as shown in
Fig. 5.2, indicating that D_L/u is independent of u. Since the
values of H and K can also be measured by the pulse response
experiment (Chaps. 3 and 4), the D_L/u values are calculated
from the slope in Fig. 5.2 as

$$\frac{D_L}{u} = \frac{slope}{(1 + HK)^2} \tag{5.10}$$

When sufficiently large molecules with K = 0, such as blue
dextran (M_w = 2,000,000) or dextran T-2000 are used as a
sample, the D_L/u value agrees with the slope.

When an ion exchanger is positively charged, blue dextran
should not be used since it is adsorbed quite strongly to the ion
exchanger. In the general case that D_L/u depends on u, the
D_L value can also be obtained from Eq. (5.9) or (5.10). Figure
5.3 shows some data on the D_L/u values obtained by the pulse
response method with nonionic gels (Sephadex gels) (Nakanishi

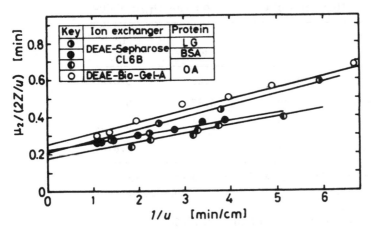

Fig. 5.2 Experimental relations between second central moment and $1/u$ for various ion-exchange columns. Elution buffer, 14 mM Tris-HCl buffer (pH 7.9) containing 0.5 M NaCl. At this ionic strength, the agarose ion-exchange gels acted as gel filtration chromatographic media. Column size, $1.5^{\phi} \times 30$ cm. LG, BSA, and OA are β-lactoglobulin, bovine serum albumin, and ovalbumin. Experiments were carried out at 20°C. (From Yamamoto et al., 1983b.)

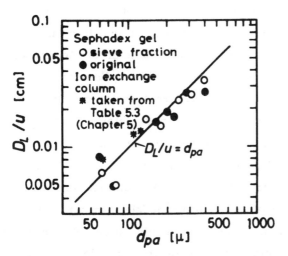

Fig. 5.3 Dependence of D_L/u on d_{pa} for various Sephadex and ion-exchange gels. (From Nakanishi et al., 1983.)

Table 5.1 Effect of Column Size on the D_L/u Value

Column size (cm)	ε	K	D_L/u^a (cm)
1.05 × 55.0	0.41	0.68	0.073
1.50 × 55.0	0.40	0.67	0.0172
2.95 × 55.0	0.42	0.72	0.0355
1.50 × 30.0	0.41	0.71	0.0157

[a]Measured at 20°C with NaCl as a sample. Sephadex G-25 medium (d_p = 164 μm) was employed as a
packing material.
Source: From Nakanishi et al. (1977b).

et al., 1983). As seen in Fig. 5.3, the D_L/u values are represented by the relation $D_L/u = d_p$ or Pe = 1.0. However, this
relation was confirmed only for columns with a diameter of 1.5
cm. The D_L/u value more or less depends on the column diameter. Table 5.1 (Nakanishi et al., 1977b) shows the effect of
column diameter on D_L/u values. The highest value is obtained
with a 1.0 cm diameter column, which may be due to a large
wall effect because of a low ratio of the diameter to the particle
size. The D_L/u value obtained with a 2.95 cm diameter column
is a little higher than that with a 1.5 cm diameter column. Although in Table 5.1 the effect of column diameter on the D_L/u
value is shown, this effect should actually be examined in terms
of the ratio of column diameter to gel particle size. Schwartz
and Smith (1953) measured the velocity distribution across a
fixed bed as a function of the column-particle diameter ratio.
They found that the column diameter should be 50—100 times the
particle diameter to obtain a practically uniform velocity profile
across the cross section of a column.

Table 5.2   Effect of Viscosity on the D_L/u Value

Eluant	Viscosity (cP)	ε	K	D_L/u[a] (cm)
5/1000 M NaCl	1.00	0.41	0.71	0.0157
+10% glycerol	1.30	0.40	0.72	0.0201
+20% glycerol	1.80	0.40	0.72	0.0238
+30% glycerol	2.50	0.40	0.72	0.0261

[a]Measured at 20°C with NaCl as a sample. A Sephadex G-25
medium column (particle diameter = 164 μm, 1.5 cm diameter and
55 cm length) was used.
Source: From Nakanishi et al. (1977b).

The D_L/u value may be also influenced by the viscosity of
the elution buffer, which may affect the nonuniformity of the
flow.   Table 5.2 shows the effect of viscosity on the D_L/u
value, by changing visocisty with glycerol. By increasing vis-
oosity by a faotor of 2.5, the D_L/u value inoroases by about
70%, as shown in Table 5.2. We have also examined the effect
of particle size distribution on the D_L/u value using nonionic
cross-linked dextran gels (Sephadex). In Fig. 5.3, closed
circles show the data obtained from a column with original gels
with a large particle size distribution, and open circles, those
with sieved gels. A mean gel diameter d_{pa} is calculated with
the relationship

$$d_{pa} = \left[\frac{\int_0^\infty d_p^5 f(d_p) \, d(d_p)}{\int_0^\infty d_p^3 f(d_p) \, d(d_p)} \right]^{1/2} \tag{5.11}$$

As seen in Fig. 5.3, there is no appreciable difference in these
D_L/u values, indicating that the effect of particle size distribu-

Table 5.3 Values of D_L/u for Ion-Exchange Columns

Ion exchanger[a]	d_{pa}[b] (μm)	Sample[c]	D_L/u[d] (cm)
DEAE-Sepharose CL6B [145/200]	110	LGA	0.0128
		OA	0.0113
		BSA	0.0126
DEAE-Bio-Gel A [145/200]	123	OA	0.0133
DEAE-Toyopearl 650 [200/250]	62	OA	0.0081

[a]The numeral in the bracket represents the range of sieving, e.g., 145/200 means a fraction between 145 and 200 mesh sizes separated from original gels.
[b]The value of d_{pa} was measured for swollen gels at I = 0.11 and pH 7.9 and calculated according to Eq. (5.11).
[c]The following abbreviations are used for proteins: bovine serum albumin, BSA; ovalbumin, OA; β-lactoglobulin A, LGA.
[d]Measured at 20°C, pH 7.9, and ionic strength 0.51 M where the ion exchangers act only as gel filtration chromatographic media. The column size was $1.5^\phi \times 30$ cm.
Source: From Yamamoto et al., 1983b.

tion is small. Table 5.3 shows the D_L/u data obtained with ion exchange columns. As seen from Fig. 5.3, the data for the ion exchangers also follow the relation $D_L/u = d_p$.

We have shown here some data on the D_L/u values obtained from columns of several kinds of nonionic gels and ion exchangers and the factors affecting these values. D_L/u values depend on column size and viscosity, but less on particle size distribution. Although it is necessary to examine these effects in more detail, we may roughly assume $D_L/u = d_p \sim 2d_p$; in other words, 0.5–1 can be used as the Pe number under the usual conditions of ion-exchange chromatography of proteins.

6

Experimental Methods and Apparatus

In this chapter, the apparatus for ion-exchange chromatography
(IEC) of proteins and such methods as the packing of the column,
the introduction of a sample, and the elution are briefly described. A large amount of important information on cellulosic
ion exchangers is found in Peterson (1970). A book by Scopes
(1982) and booklet distributed by a manufacturer of ion-exchange
gels (Pharmacia Fine Chemicals, 1987) contain a useful description of the experimental method. The experimental method and
the apparatus described in the references on high-performance
liquid chromatography (HPLC) (for example, Snyder and Kirkland, 1974) are basically the same as those for high-performance
IEC (HPIEC) of proteins. A useful short review for the care
of HPLC columns has been reported by Wehr (1984), in which
troubleshooting for the column is also briefly summarized.

A general flow sheet for the preparation and operation of IEC
of proteins is given in Fig. 6.1, and Fig. 6.2 schematically
illustrates the operation procedure of the ion-exchange column.
We first review several pieces of apparatus usually employed in

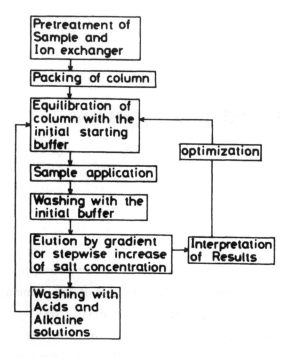

Fig. 6.1 Flow sheet for the preparation and operation of IEC of proteins.

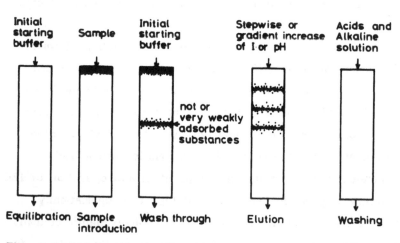

Fig. 6.2 Schematic drawing of the operation procedures of ion-exchange columns.

IEC and then describe the preparation and operation of IEC of
proteins according to the flow sheet shown in Fig. 6.1.

6.1 APPARATUS

A typical experimental setup for the separation and purification
of proteins by IEC is shown schematically in Fig. 6.3. Of
course, we can operate without some of the equipment shown
in the figure or replace it by much simpler and cheaper equipment. In contrast, several pieces of apparatus can be added to
this setup, such as a microcomputer for automatic data acquisition
and control of elution schedules. The HPIEC system is basically
similar to that shown in Fig. 6.3, although the individual parts
are somewhat different from those in low-pressure IEC (LPIEC).

Pumps

Although the column can be operated without a pump if the
pressure drop is low (less than 0.1 atm), we recommend use of
the pump. It gives better reproducibility and reduces tedious
procedures to maintain the constant flow rate by a gravity force.
In addition, as described in Chap. 7, a continuing development
of ion exchangers is toward the reduction of the particle size

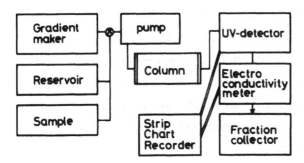

Fig. 6.3 Block diagram of a typical experimental setup for IEC
of proteins.

d_p. This results in HPIEC and medium-performance IEC (MPIEC). As described in Chap. 7, the pressure drop at the same flow rate and column length increases with d_p^{-2}. This increase in the pressure drop makes the operation of the column without the pump difficult. Peristaltic pumps (tube pumps) are sufficient for LPIEC and MPIEC of $d_p > 30$ μm. Care must be taken to avoid the deformation and breakage of the tubes, which cause not only a variation in flow rate but also leakage of the liquids. For continuous and/or large-scale operation, peristaltic pumps are not suited and other types of pumps, such as syringe pumps and reciprocating piston pumps, should be used. Since a pressure force higher than 2 atm is required for the HPIEC column, a high-pressure pump without pulsation must be used.

Detectors

For the continuous detection of proteins in the effluent, ultraviolet (UV) detectors are commonly used. Many types of flow cells for the UV detector are now commercially available. The cell with a short path length is useful for preparative purposes; highly sensitive detection requires a long path length and small cell volume. Many attempts have been made in the development of selective detectors that are useful for the detection of several enzymes (Regnier and Gooding, 1980).

Electroconductivity detectors are useful for the continuous measurement of salt concentrations in the gradient and stepwise elution by salts. It is also a good practice to check the equilibration of columns with this detector. Although pH is another important factor that must be monitored, a pH detector is not yet a common apparatus.

Generally, it is desirable that the detector have a high sensitivity with low noise, linearity, and a quick response.

Columns

In many laboratories, bore glass or acrylic tubes at the bottom of which a glass filter or glass wool is mounted are often employed as a column tube for low-pressure IEC. This type of column is often called an open-ended column or simply an open column. However, to obtain good separation with a column packed with smaller ion exchangers, a column tube with end fittings at both ends, called a closed column, should be used. A dead volume contained in the end fittings must be reduced as much as possible so that the contribution to the total zone spreading of nonchromatographic zone spreading in the dead volumes becomes negligible. Most commercial column tubes for low-pressure liquid chromatography are constructed of transparent materials, such as glass and acrylic resins. Such columns are very useful for the IEC of proteins, since we can observe the movement of the zones by using nonionic colored substances and can check the fouling of ion exchangers. A narrow-diameter stainless steel column tube is usually employed for HPIEC since high-pressure forces are applied to the column. The narrow-diameter column is resistant to such high-pressure forces because the fraction of the gel beads contacting with the column tube wall becomes larger with a decrease in the column diameter. Since we cannot directly observe change in ion exchangers packed in the stainless steel column, care must be taken with the excess pressure forces, which cause a vacant dead space between the top of the gel bed and the end fitting. This markedly lowers column efficiency.

Gradient Makers

Several types of automatic gradient makers are now commercially available. Some of them are reviewed by Snyder and

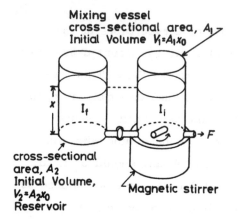

Fig. 6.4 Simple device for making a linear gradient with two
vessels with an equal cross-sectional area.

Kirkland (1974). However, these devices are very expensive.
A simple device that uses two vessels of equal cross-sectional
area, shown in Fig. 6.4 (Bock and Ling, 1954; Drake, 1955;
Schwab et al., 1957), has been successfully employed for making
a linear gradient in the laboratory-scale IEC of proteins. The
slope of the linear gradient obtained with this device is given by

$$I = I_i + gFt \qquad\qquad (6.1)$$

where $g = (I_f - I_i)/(V_1 + V_2) = (I_f - I_i)/V_G$ and $V_G = V_1 + V_2$.

I_i = ionic strength of the initial starting buffer solution contained in the mixing vessel

I_f = ionic strength of the final solution in the reservoir

V_1, V_2 = volume of buffer solution initially filled in the mixing vessels and in the reservoir; in this case they
are equal; that is $V_1 = V_2 = V_G/2$

F = volumetric flow rate

Drake (1955) and Bock and Ling (1954) have presented equations for various shapes of gradients produced by simple devices consisting of one or two vessels. If the cross-sectional areas of two vessels are not the same, the shape of gradient can be given by Eq. (6.2):

$$I = I_f + (I_i - I_f) \left(1 - \frac{Ft}{V_G}\right)^h \qquad (6.2)$$

where $h = A_2/A_1$. When $A_1 = A_2$, Eq. (6.2) reduces to Eq. (6.1). When only one mixing vessel with a constant volume is used, the resulting shape of the gradient is given by

$$I = I_f - (I_f - I_i)e^{-Ft/V_G} \qquad (6.3)$$

When Ft/V_G is small, this equation is approximated as

$$I = I_i + (I_f - I_i) \frac{Ft}{V_G} \qquad (6.3')$$

Several results calculated by these three equations are shown in Fig. 6.5. It can be seen that the initial part of the exponential gradient represented by Eq. (6.3) can be approximated by a linear gradient of the slope $= (I_f - I_i)/\overline{V}_G$. Several other types of gradient devices can be found in the literature (Bock and Ling, 1954; Drake, 1955; Schwab et al., 1957).

The linear gradient device shown in Fig. 6.4 can be easily made in the laboratory. It should be kept in mind, however, that such a device is not suitable when the densities of the initial and final solution are quite different. When the device is handmade, care must be taken against air bubbles trapped in a tube connecting the two vessels. The trapped air bubbles disturb liquid flow from the reservoir to the mixing vessel, and

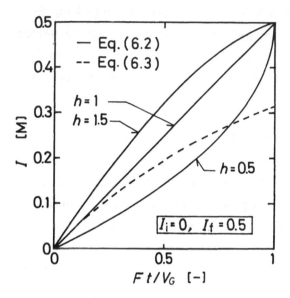

Fig. 6.5 Gradient shapes calculated by Eqs. (6.2) and (6.3).

hence an expected shape of the gradient may not be obtained. Gradient shapes given by Eqs. (6.1) through (6.3) can also be obtained with the aid of two pumps or a two- (or three-) channel pump, as shown schematically in Fig. 6.6. In this case, V_G and h in Eq. (6.2) are given by $V_G = V_1/(1 - F_2/F)$ and $h = F_2/(F - F_2)$. This type of the gradient maker is used in some commercial HPIEC systems (Snyder and Kirkland, 1974). In some automatic gradient makers, we must set the gradient time t_G rather than the gradient volume V_G. Consequently, t_G frequently appears in the reference as a variable in HPLC. t_G is related to V_G as $V_G = t_G F$, where F is the volumetric flow rate. Problems in making a gradient using the conventional (commercial) apparatus were discussed by Kaminski et al. (1984). Van der Zee and Welling (1984) have presented a method for

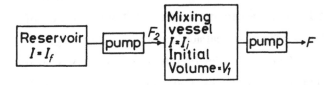

Fig. 6.6 Block diagram for making a linear gradient with two pumps.

making a gradient that uses a personal computer with a three-way solenoid valve. An interesting device for making a linear gradient that uses a gas bomb instead of pumps was employed for preparative protein separation on DEAE-cellulose ion-exchange columns (Bergenhem et al., 1985).

6.2 PRETREATMENT

Several pretreatments are required for both ion exchangers and samples in order to obtain successful results by IEC. When ion-exchange gels are supplied as dried powders, they must be brought in the swollen state, according to the manufacturer's information. When ion-exchange gels are distributed in the wet form, this procedure can be skipped. The next step is to make a gel slurry appropriate for packing columns. Equilibration of ion-exchange gels with the initial buffer may also be performed in this step by several decantations. Very fine particles in the slurry are also removed by decantation. Washing with acidic and alkaline solutions is not necessary for most ion-exchange gels, except cellulosic ion exchangers. Pretreatments of cellulosic ion exchangers are described by Peterson (1970).

A sample to be introduced to ion-exchange columns should also be equilibrated with the initial starting buffer. For this purpose, dialysis with a semipermeable tube or gel filtration chromatography (GFC) is frequently used. A large number of useful suggestions for the pretreatment of the sample is given in Peterson (1970). Several useful charts for the prediction of operating conditions of desalting by GFC have been presented by Nakanishi et al. (1979). An alternative simple and rapid technique is dilution. When the initial condition is chosen so that the protein adsorption capacity is high, the degree of separation of a diluted sample is equal to or often superior to that of the original sample, as shown in Chap. 8. However, this requires knowledge of the adsorption capacity as a function of ionic strength and pH. In addition, some enzymes are unstable at low concentrations.

6.3 PREPARATION OF ION-EXCHANGE GEL COLUMNS

Most ion-exchange gels are usually packed by a slurry packing method. The concentration of a slurry should be low enough for air bubbles to escape freely. However, a very thin slurry may cause unevenly packed beds owing to convection and size segregation of particles during packing. A slurry containing approximately 30—40% wet gels may be suitable for most soft and semirigid ion-exchange gels (Kato et al., 1981a,b; Pharmacia Fine Chemicals, 1987). However, our experience indicates that the optimal concentration of the slurry is not constant but variable with the properties of ion-exchange gels, such as the particles diameter and the density of the gel. We should thus adopt the concentration recommended by the manufacturer. If this is not available, several attempts with different slurry concentrations are helpful for determining the suitable slurry concentration for the packing procedure.

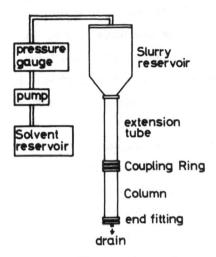

Fig. 6.7   Typical setup for packing columns.

Next we must choose a method for packing columns and a
setup.   The most useful setup is sketched in Fig. 6.7.   It is
desirable that the slurry is added in one operation to obtain
uniform packing.   For soft compressible ion-exchange gels, the
pressure drop must be lower than the value at which the maximum flow rate can be obtained, as discussed in Chap. 7.   In
contrast, several researchers (Edwards and Helft, 1970; Kato
et al., 1981a,b) have shown that the column efficiency of a tight-
ly packed rigid gel column is higher than that of a loosely packed
column.   We obtained similar findings for the column efficiency
of gel filtration columns of rigid gels.   When the packing velocity was low, the elution curve became asymmetric and had a
tail.   Compression of this gel column by supplying an eluent at
high flow rates improved the poor column efficiency and the
elution curve became symmetrical.   A possible reason for this
result is that shortcut paths or dead spaces, which cause

channeling and uneven flow, tend to be formed during the
packing of such rigid gels when the pressure force is low. A
disadvantage of this packing method is that an extension tube
and/or a reservoir is required, whose cost is not negligible. In
addition, it is difficult in the laboratory to pack larger columns
by this method.

Another method that does not require an extension tube is
as follows. The slurry is first poured into the column tube
until it fills one-half to two-thirds of the column height. Then,
a liquid is pumped continuously from the outlet of the column
either with the pump or by a gravity force. Fines are removed
from the supernatant of the slurry while the bed is forming.
The slurry is added several times until the desired height of
the gel bed is completed. Soft ion-exchange gels can be packed
uniformly by this method.

A packing method that utilizes one characteristic of gels
that shrink in organic solvents or salt solutions of high concentration may be advantageous in packing a flat thin column of
large diameter. Gels shrunk with such solutions are packed in
the column, and then they are allowed to swell by replacing the
solvent.

After the gel bed of desired length is completed by either
of the preceding three packing method, a buffer solution of
several column volumes is supplied to stabilize the gel bed.
When we use the column with end fittings at both ends, an end
fitting must be mounted at the top of the column. This procedure sometimes requires empirical skill. However, this so-
called closed column must be prepared for medium- and high-performance chromatography. Even for low-pressure chromatography
with soft gels, the closed column is useful for repeated use and

for upward flow operation. Care must be taken against fouling
of the gel bed since the topmost part of the gel bed cannot be
observed in several commercial column tubes, although the column
tube consists of glass. On the other hand, open columns (open-
ended columns) in which there is a vacant space between the top
of the gel bed and the inlet of the column are sometimes useful
since the fouling can be easily checked and removed. In addition, a sample can be applied without special injection device
simply by the use of a pipette, as shown in Sec. 6.4. The open
column is recommended for the small column packed with soft
ion-exchange gels. To keep the surface at the top of the gel
bed flat, a filter paper should be placed on it (see Fig. 6.9).

We must pay special attention to the equilibration of the ion-
exchange column to obtain successful and reproducible results.
Usually, equilibration of the column with the initial starting buffer of several column volumes is sufficient (Kato et al., 1982a).
However, in the case of low ionic strengths and/or after washing of the column with acids or alkaline solutions, the volume
required for the equilibration becomes large. It is not unusual
that buffer of more than 20 column volumes is required for
equilibration. In such a case, it is recommended that the column be titrated with appropriate acidic or alkaline solutions so
that the pH approaches that of the initial buffer, as Peterson
(1970) stated. It is advisable that the pH and ionic strength of
the effluent be checked whether or not the column is equilibrated.

6.4 METHOD FOR SAMPLE INTRODUCTION

As already pointed out by several researchers (Morris and
Morris, 1964; Coq et al., 1979; Kaminiski et al., 1982), the

method for sample introduction must be examined carefully because elution curves from tilted zones due to poor sample introduction, shown in Fig. 6.9, may not be distinguished from those
of horizontal zones. When isocratic elution chromatography is
employed to obtain a fine separation, poor sample introduction
becomes a serious factor in lowering column efficiency. On the
other hand, in gradient elution, the zone sharpening effect described theoretically in Sec. 2.1 and experimentally in Sec. 8.1
can improve the tilted zone during elution; such improvement
cannot be expected in isocratic elution. Although sample volume
is one of the dominating factors limiting separation efficiency in
isocratic elution, it does not significantly affect separation efficiency in gradient elution when appropriate initial starting conditions are chosen.

A six-port injection valve is most frequently employed as a
device for introducing a small amount of the sample to a closed
column (see Fig. 6.8). Usually, a sample solution is filled in a
loop, the volume of which is adjusted as a sample volume. However, for the introduction of a large sample volume, a very long

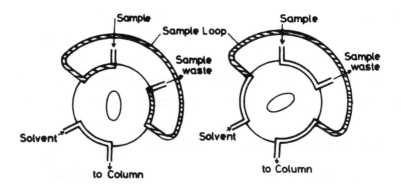

Fig. 6.8 Mechanism of a six-port injection valve.

loop is required. For example, if the diameter of a tube used
for the sample loop is 0.1 cm, the tube, the length of which is
nearly 13 m, is required to inject a sample volume of 10 mL. A
convenient device for the introduction of a large sample volume,
which can be used instead of the sample loop, is commercially
available under the trade name Super Loop from Pharmacia Fine
Chemicals. However, the introduction of a large volume of a
sample solution directly to the column with a pump usually gives
good results when the initial starting conditions are such that
a large amount of protein in the sample solution can be adsorbed
to an ion-exchange column as a small zone from the top of the
column.

A sample can be introduced to an open column simply by the
use of a pipette. A technique for this method is shown schematically in Fig. 6.9. When soft ion-exchange gels are used,
care must be taken in the speed of dropping the sample from

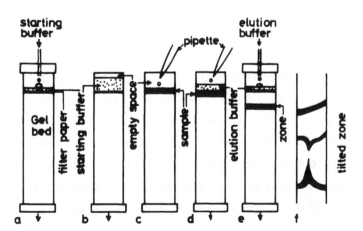

Fig. 6.9 Method for sample introduction to an open-ended column. Several typical tilted zones due to poor sample introduction are illustrated schematically in column f.

the pipette. If the speed is too great, the bed is deformed by the flow of the sample solution from the pipette.

We recommend that the packing of a column be checked with nonionic colored substances. Blue dextran T-2000, which is commonly employed for determining the void fraction of a column, cannot be used for anion exchangers because it is strongly adsorbed to the ion exchangers and is not easily removed. When observed zones with such a colored substance are not horizontal but tilted, as shown in Fig. 6.9, the column must be repacked. A prepacked HPIEC column is usually tested by the manufacturer with a test sample before the shipment, and the data accompany the column. It is therefore recommended that the efficiency of the HPIEC column be occasionally measured against the test sample in order to monitor the deterioration of the column.

6.5 METHOD FOR ELUTION

After sample introduction, the column is usually washed with the starting buffer of one or several column volumes to purge components that are not adsorbed or are only very weakly adsorbed (see Fig. 6.2). In the isocratic elution method, of course, substances are separated in this step. For nonisocratic elution, such as gradient and stepwise elution, the next step is elution by a continuous or discontinuous change in the elution buffer. In gradient elution, the gradient of the elution buffer produced by the devices described in Sec. 6.1 is applied to the column. In stepwise elution, we must change the elution buffer according to the elution schedule either manually or with the aid of an automatic valve. Effluent from the column is monitored continuously with a UV or another type of detector or is collected with an automatic fraction collector. Protein concentrations and enzymatic activities in the fractions are assayed by suitable methods for the relevant conditions.

6.6 REGENERATION OF THE COLUMN

If we want to use the column repeatedly, it must be washed and re-equilibrated. However, when the top part of the column is markedly fouled and/or the column shrinks significantly, it is better to unpack the column and, if necessary, dispose of the fouled ion exchangers. The ion exchangers are washed with acidic and alkaline solutions. Usually, 0.1 M sodium hydroxide solution and 0.1 M hydrochloric acid solution are used. The column is then re-equilibrated with the initial starting buffer. If the fouling is not severe and/or the ion exchangers are not allowed to be present in such strong acidic or alkaline solutions, the column is washed with solutions that are less acidic or alkaline than those just described. In general, we should follow the washing procedure recommended by the manufacturer. Maintenance of the ion-exchange column is discussed in detail in Sec. 9.3.

6.7 CONCENTRATION OF FRACTIONS

Concentration of the fractions is needed for their preservation and to perform further purification steps. Various methods are available for this purpose, such as ultrafiltration, salting out, dehydration using hydrophilic gels of a large water regain, and freeze drying.

Ultrafiltration is very efficient, and the time required for protein concentration is very short compared with that in other techniques. In addition, this can be operated continuously. A hazard involved in this method is inactivation of enzymes due to shear stress, as pointed by Charm and Matteo (1971). On the basis of the same principle, protein solutions contained in a semipermeable tube can be simple concentrated according to the procedure described by Peterson (1970).

It is known that proteins precipitate in the presence of salts, such as ammonium sulfate and sodium sulfates. This phenomenon is called salting out. In addition to such salts, organic solvents, such as ethanol, and polymers, such as polyethylene glycol, can be employed as precipitants. With organic solvents, operation must be carried out under 0°C so that protein denaturation does not occur. Charm and Matteo (1971) and Scopes (1982) have described important information on the protein precipitation technique.

Protein solutions can be concentrated when they are introduced to a semipermeable tube in contact with hydrophilic gels of a high water regain, such as Sephadex G-100 or G-200. This procedure is very simple and does not require special apparatus. Moreover, since the concentration process is carried out under mild conditions, we may expect a very low degree of protein denaturation.

Another efficient method for protein concentration is evaporation of solvent from the solution to the surrounding media. It is known that most enzymes become stable with a decreasing moisture content, and inactivation of enzymes in dehydration processes occurs at the stage when both the temperature and the moisture content of the material to be dehydrated are high (Luyben et al., 1982; Yamamoto et al., 1985). Therefore, in various dehydration methods, the temperature must be kept low until the moisture content decreases to a certain low level where enzymes are stable. Care must also be taken in the change in pH and/or ionic strength during the dehydration processes, since the stability of enzymes usually strongly depends on such factors. A very simple method belonging to this type of concentration method can be carried out with a semipermeable tube. We can concentrate protein solutions contained in a semipermeable tube by blowing air through the tube. The most frequently employed method is freeze drying becuase it is operated at low temperature. However, we should bear in mind that some enzymes are unstable when they are frozen.

Ion-exchange chromatography itself is also regarded as one of the concentration methods for proteins. As described in the following sections, most ion-exchange gels can adsorb a large amount of protein if proper initial conditions are selected. The elution of adsorbed proteins with a high ionic strength buffer then yields a very concentrated fraction of the proteins. An example of the application of IEC to the concentration of proteins is given in Sec. 8.5 (see Table 8.2). This method can be carried out without using a column and is suited for large-scale concentration, such as the concentration of proteins from cultivation media.

Each of the preceding methods has both advantages and disadvantages. Therefore, we should select the concentration method suitable for particular conditions.

6.8 OTHER SUBJECTS

In addition to the subjects discussed thus far, several other subjects must be examined carefully. The preparation of buffer

systems is one such subject. Bjork (1959) pointed out that
carbon dioxide dissolved in the buffer solution caused the anomalous elution behavior of chloride concentrations and pH. Moreover, gases dissolved in the buffer may be a trigger for such
problems as air bubbles trapped in the flow cell of a UV detector. We therefore recommend that a buffer solution be degased
under vacuum before use. An automatic degaser is also commercially available. A buffer solution has its maximum buffer
action at a pH near its respective pK. Therefore, the choice
of buffer solution depends on the pH to be employed. The references on the buffer systems (Perrin and Dempsey, 1974; Peterson, 1970; Scopes, 1982; Blanchard, 1984) are helpful in select-
ing a buffer system suitable for different chromatographic conditions. The pH and ionic strength I of the starting and elution
buffer should also be such that proteins or enzymes of interest
are stable in the buffer. Several materials, such as mercaptoeth-
anol and glycerol, which may stabilize proteins, are added to the
buffer if necessary. When the ionic strength is very low, a
protein may be adsorbed so strongly to ion exchangers that its
conformation may be destroyed, leading to denaturation. In
addition, salts are generally stabilizing reagents for proteins.
The initial ionic strength should therefore not be too low. The
buffer solution as well as the sample should be filtered with a
Micropore filter (at least 0.45 μm and preferably 0.22 μm) before use to prevent the ion-exchange columns from fouling.
For this reason, the water used must also be of high purity.
Ganzi (1984) has summarized the preparation of high-purity
laboratory water. Nonchromatographic zone spreading, such as
the zone spreading in the column dead volume and the column

accessories, must be eliminated. Such nonchromatographic zone spreading was investigated previously (Arnold et al., 1985b; Hupe et al., 1984; Janson and Hedman, 1982).

7
Ion Exchangers

Since the introduction of ion-exchange cellulose by Peterson and Sober in 1956, various types of ion exchangers (packings) have been developed of ion exchangers and their matrices was given in Sec. 1.3.

The following properties are required by ion exchangers for protein separation.

1. The ion-exchange capacity must be high enough for a large amount of proteins to be adsorbed.
2. Ion-exchange groups and the gel matrix must be stable enough to withstand washing with acids or alkaline solutions.

The following are required not only for ion exchangers but also for packing materials used in other chromatographic methods, such as gel filtration chromatography and affinity chromatography.

3. The matrix should be hydrophilic, insoluble, and stable.
4. Any nonspecific adsorption interaction is undesirable and should be eliminated.

5. Because proteins are large moelcules, an ion-exchange matrix must have large pores that are accessible to the proteins.
6. The shape of an ion exchanger should be spherical and the distribution of the particle diameter as small as possible.
7. Ion exchangers should be mechanically stable, that is, neither deformed nor compressed by pressure.
8. They should not shrink or swell with a change in pH or salt concentration.
9. The cost should not be high and they should be able to withstand repeated use.

Properties 2 and 7 through 9 are especially important for a continuous use or a large-scale separation. In addition to these properties, it is useful if ion exchangers are autoclavable and can be used in the presence of denaturation reagents, such as urea and guanidine HCl. Interlot variation among ion exchangers must also be low. In actuality, it is not easy to produce an ion exchanger that fulfills all these requirements. For example, in general, an increase in pore size decreases the mechanical stability of the hydrophilic gel bead. As the ion-exchange capacity of such gels is increased, the gels shrink or swell markedly with a change of ionic strength and/or pH. In the past few years, numerous efforts have been devoted to the development or improvement of ion exchangers. Current ion exchangers are superior to conventional soft ion-exchange gels. Typical efficient packings are very small and rigid gels used in high-performance ion-exchange chromatography (HPIEC). These packing materials were made possible by recent advances in high-performance liquid chromatography (HPLC). Although it is difficult to define conventional low-pressure IEC (LPIEC), medium-performance IEC (MPIEC), and HPIEC in terms of the particle diameter d_p, we have made a tentative correlation between d_p and the separation time for these three types of IEC columns (Fig. 7.1).

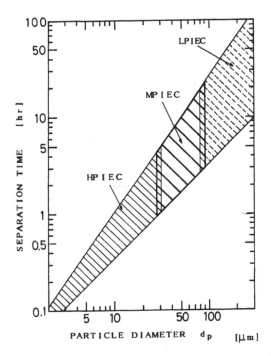

Fig. 7.1 Relation between separation time and particle diameter. The shaded area indicates a tentative correlation between the separation time and the particle diameter. Although there is no confirmed definition of LPIEC, MPIEC, and HPIEC, we classify these in terms of the particle diameter according to the properties of commercially available ion-exchange gels.

Typical experimental elution curves of a crude enzyme on MPIEC and HPIEC columns are shown in Fig. 7.2. The elution behavior of HPIEC and MPIEC is similar, although many small peaks are found in HPIEC that cannot be resolved well in MPIEC. We see that the resolution and the speed of separation become higher with a decrease in d_p.

The properties of several commercial LPIEC and MPIEC packings for protein separation have been compiled by Janson and Hedman (1982), Kato et al. (1982c), and Unger and Janzen

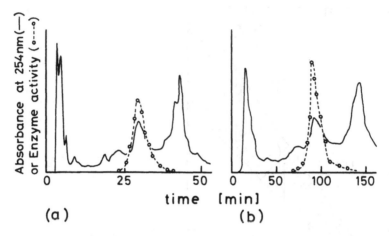

Fig. 7.2 Elution curves of crude β-galactosidase with (a) high- and (b) medium-performance ion-exchange chromatography. The sample used is crude β-galactosidase from *Aspergillus oryzae*. The linear gradient elution was performed by increasing NaCl concentration in 14 mM Tris-HCl buffer (ionic strength = 0.01, pH 7.7) from 0.03 M at 25°C. (a) High-performance column, TSK gel DEAE 5PW (d_p = 10 μm, d_c = 0.75 cm, Z = 7.5 cm). The slope of the gradient g = 0.0047 M/mL; the flow rate F = 1 mL/min; sample, 1%, 1 mL. (b) Medium-performance column, DEAE-Toyopearl 650S (d_p = 40 μm, d_c = 1.6 cm, Z = 15 cm): g = 6 × 10^{-4} M/mL; F = 2 mL/min; sample, 1%, 10 mL. The dotted and solid curves represent the activity of β-galactosidase (arbitrary units) and the absorbance at 254 nm (UV detector output, arbitrary units), respectively. (From Yamamoto et al., 1987b, by permission of the authors and Elsevier Publishing Company.)

(1986). HPIEC packings and columns are treated in the reviews by Majors (1977), Wood et al. (1980), Regnier (1984), and Unger and Janzen (1986). The last reference is focused on preparative LC. Table 7.1 lists the properties of LPIEC and MPIEC packings taken from Janson and Hedman (1982) and Kato et al. (1982c). These gels are usually packed into the column tube by the end user. On the other hand, HPIEC columns are available as pre-

Table 7.1 Properties of Several Anion Exchangers for Protein Separation

Ion exchanger: DEAE with[a]	Exchange capacity (meq/mL)		Protein adsorption capacity (mg/mL)		Particle diameter (μm)[f]	Ion-exchange matrix
	pH 7.0[b]	I = 0.1[c]	pH 8.3[d]	pH 8.0, I = 0.01[e]		
Sephadex A-25	0.354	0.5	22	70	150	Dextran gel
Sephadex A-50	0.030	0.175	78	250	250	Dextran gel
Sepharose CL-6B	0.114	0.15	97	100	100	Cross-linked agarose gel
Bio-Gel A	0.013		28		100	Cross-linked agarose gel
Toyopearl 650	0.108		26		40, 60, 90	Hydrophilic vinyl polymer gel
Trisacryl M	0.175	0.3	73	85	40–80[h]	Acrylic copolymers
Cellulose DE-23	0.108	1.0[g]	58			Fibrous cellulose
Cellulose DE-52	0.118	1.0[g]	116			Microgranular cellulose
Sephacel	0.093	1.4[g]	89		40–160[h]	Beaded cellulose

[a]DEAE, diethylaminoethyl group. The manufacturers of the ion exchangers are Pharmacia Fine Chemicals AB, Sephadex, Sepharose, and Sephacel; Bio-Rad Laboratories, Bio-Gel A; LKB, Trisacryl M; Whatman Biochemicals Ltd., Cellulose DE; Tosoh Corp., Toyopearl.
[b]From Kato et al. (1982c).
[c]From Janson and Hedman (1982).
[d]Values for bovine serum albumin in Tris-HCl buffer. From Kato et al. (1982c).
[e]Values from hemoglobin in Tris-HCl buffer. From Janson and Hedman (1982).
[f]The approximate value of the average particle diameter is cited from Yamamoto et al. (1983b, 1987d).
[g]Exchange capacities are given in meq/g.
[h]From the leaflet distributed by the manufacturer.

packed columns. Data for commercial HPIEC columns taken from
the leaflet distributed by the manufacturers are shown in Table
7.2. Similar tables containing more detailed information are
found in Regnier (1984) and Unger and Janzen (1986). Many
commercial HPIEC packing materials are silica-based packings,
which are not stable at high pH. The dimensions of HPIEC
columns are usually 0.5–0.8 cm in diameter and 5–10 cm in
length. Recently, several preparative IEC columns up to 5.5
cm in diameter have been commercially available (Nakamura et al.,
1985) and have been applied to protein fractionation (Brewer
et al., 1986). It follows from these tables that most ion-exchange gels have an ion-exchange capacity of 0.1–0.2 meq/mL.
It should be noted that the ion-exchange capacity usually
changes with pH, as shown in Fig. 7.3. The effect of pH and
ionic strength on the change in several LPIEC and MPIEC column
bed volumes is shown in Fig. 7.4. The bed of some IEC columns
shrinks with increasing pH and/or ionic strength (salt concentration); others do not change appreciable. The latter columns
are useful for repeated use and for large-scale separation since
they can be washed and re-equilibrated without repacking procedures.

   Knowledge of the relation between pressure drop ΔP and
flow rate is also important, especially in large-scale or preparative IEC, as discussed in Chap. 9. The well-known Kozeny-
Carman equation relates ΔP to the particle diameter d_p, the
flow rate, the viscosity of the solvent (buffer solutions) η, the
column length Z, and the void fraction of the column ε (Carman,
1937):

Table 7.2 Properties of Several High-Performance Anion-Exchange Chromatography Columns for Protein Separation

Column	d_p (μm)	Ion-exchange capacity (mEq/mL)	Ion-exchange matrix	Supplier
Mono Q	10	0.28–0.36	Hydrophilic polymer	Pharmacia
TSK gel DEAE-5PW	10	0.1	Hydrophilic polymer	Tosoh
TSK gel DEAE-3SW	10	0.13	Silica gel	Tosoh
TSK gel DEAE-NPR	2.5	0.13	Nonporous polymer	Tosoh
Asahipak ES-502C	9		Hydrophilic polymer	Asahi Kasei
SynChropak AX-300	5, 10		Silica gel	Syn Chrom
Nucleogen DEAE-500	10		Silica gel	Macherey-Nagel
BAKER BOND Wide-pore PEI	5		Silica gel	J. T. Baker
Shim-pack WAX-2	5		Silica gel	Shimazu

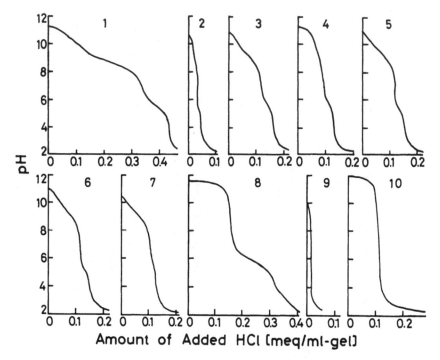

Fig. 7.3   Titration curves for several commercial ion exchangers. The numbers in the figure imply: 1. DEAE-Sephadex A-25, 2. DEAE-Sephadex A-50, 3. DEAE-Sepharose CL-6B, 4. DEAE-Sephacel, 5. DEAE-Cellulose DE52, 6. DEAE-Cellulose DE23, 7. DEAE-Cellulose, 8. DEAE-Trisacryl M, 9. DEAE-Bio-Gel A, 10. DEAE-Toyopearl 650M. Reproduced from Kato et al. (1982c) by permission of the authors and Elsevier Scientific Publishing Co.

$$\Delta P = \frac{180 \eta Z u_0 (1 - \varepsilon)^2}{d_p^2 \varepsilon^3} \tag{7.1}$$

where $u_0 = F/A_c = u\varepsilon$ = superficial velocity

   F = volumetric flow rate

   A_c = column cross-sectional area

   u = linear mobile-phase velocity

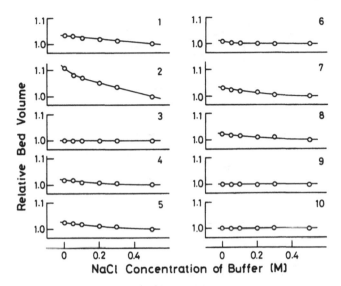

Fig. 7.4a Dependence of bed volume on NaCl concentration for several commercial ion exchangers. Buffer; 0.02 M phosphate (pH 7.5). Column size; $1.6^{\phi} \times 35$–45 cm. The numbers in the figure are the same in Fig. 7.3. Reproduced from Kato et al. (1982c) by permission of the authors and Elsevier Scientific Publishing. Co.

This equation tells us that the pressure drop increases linearly with (1) the viscosity of the solvent, (2) the column length, (3) the flow rate, and (4) the inverse of the square of the particle diameter. It is noteworthy that even for a column packed with ion exchangers of the same diameters the pressure drop varies with the void fraction. For example, lowering the void fraction from 0.40 to 0.34 brings about an approximate two-fold increase in the pressure drop. Actually, Eq. (7.1) is valid only for the rigid ion exchangers that are not deformed by pressure. When ion exchangers are soft and compressible, the flow rate u is a linear function of the pressure drop ΔP at a lower ΔP. However, as ΔP is increased, the slope of u versus ΔP

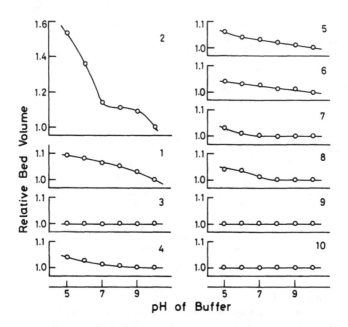

Fig. 7.4b Dependence of bed volume on pH for several commercial ion exchangers. Buffer; 0.05 M phosphate. Column
size; 1.6 × 35–45 cm. The numbers in the figure are the same
in Fig. 7.3. Reproduced from Kato et al. (1982c) by permission
of the authors and Elsevier Scientific Publishing Co.

decreases and, finally, u shows a maximum at a certain ΔP. A
further increase in ΔP does not increase u but decreases u (see
Fig. 7.5). This phenomenon is attributed to bed compression
(Janson and Hedman, 1982; 1987; Buchholz and Godelmann, 1978).
Therefore, for the use of such compressible ion exchangers we
must pay special attention to the pressure drop so that it does
not exceed the maximum pressure drop. The relation between
ΔP and u for various MPIEC columns was also measured by
Larre and Gueguen (1986).

   As explained theoretically in Chap. 2 and demonstrated experimentally in Chap. 8 (see also Fig. 7.2), HPIEC can offer a

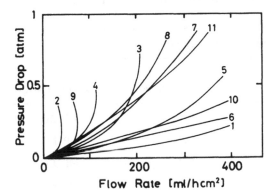

Fig. 7.5 Relationships between pressure drop and flow rate for several commercial ion exchangers. Column size; 1.6 × 15 cm. Eluent; 0.1 M NaCl solution. The numbers in the figure are the same as in Fig. 7.3. The number 11 is DEAE Toyopearl 650S. Reproduced from Kato et al. (1982b) by permission of the authors and Elsevier Scientific Publishing Co.

very high separation efficiency and a high speed of separation. We should keep in mind, however, that the pressure drop increases with the reduction in particle diameter already mentioned. Let us calculate using Eq. (7.1) the pressure drop ΔP of a column 15 cm in length packed with gels of various particle diameters. We assume that the void fraction $\epsilon = 0.4$ and the solvent viscosity $\eta = 1$ cP = 1 mPa·s. The calculated results are tabulated in Table 7.3. We see that very high pressure forces are required for a column of small particles especially those less than 20 μm in diameter. According to our experience, operation of the IEC column becomes exceptionally difficult as the pressure drop is increased to 1—1.5 atm. Under the usual operating conditions, this corresponds to a column with particles less than 20—30 μm, as shown in Table 7.3. This is one of the reasons we refer to IEC with particles less than 30

Table 7.3   Relation Between Pressure Drop and Particle
Diameter

Particle diameter (μm)	Pressure drop (atm)[a]		Ion exchange
	at u_0[b] = 1 cm/min	at u_0 = 2 cm/min	
200	0.0063	0.013	LPIEC
100	0.025	0.05	LPIEC
40	0.16	0.32	MPIEC
20	0.63	1.26	MPIEC
10	2.5	5.0	HPIEC
2.5	40.0	80.0	HPIEC

[a]In calculating pressure drop, the following values are employed:
solvent viscosity = 1; cP = 1 mPa·s; column length = 15 cm;
void fraction = 0.4.
[b]Superficial velocity = volumetric flow rate/column cross-sectional area.

μm in diameter as HPIEC. When such high-pressure forces are
required, the column tube, the column end fittings, the pipes,
and the accessories that tolerate high pressures, as well as a
high-pressure pump, must be employed. The cost of packing
materials increases drastically with a decrease in the particle
diameter d_p. For example, a reduction in d_p from 100 to 10 μm
causes a nearly fivefold increase in cost (Unger and Janzen,
1986). Packing materials with a d_p between 20 and 60 μm,
which we refer to as MPLC, are recommended for preparative
and process LC (Regnier and Gooding, 1980; Janson and Hedman, 1982; Unger and Janzen, 1986).

In contrast to gel filtration chromatography, in which suitable
gels can be easily determined in terms of their molecular sieving

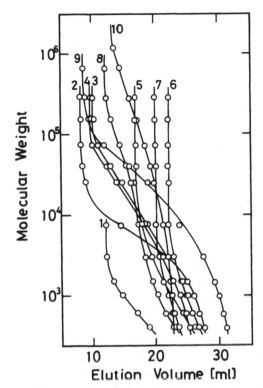

Fig. 7.6 Molecular weight calibration curves of several commercial ion exchangers for polyethylene glycols. Column size;
1.6 × 15 cm. Eluent; 0.1 M NaCl solution. Flow rate; 0.5--1
mL/min. The numbers in the figure are the same in Fig. 7.3.
Reproduced from Kato et al. (1982c) by permission of the
authors and Elsevier Scientific Publishing Co.

range, the choice of a suitable IEC column is rather complicated.

Kato et al. (1982c) measured the molecular weight calibration

curves of several commercial LPIEC and MPIEC columns. As

shown in Fig. 7.6, the shapes of the calibration curves are dif-

ferent. The elution volumes for DEAE-cellulose ion exchangers

are insensitive to molecular weight. The interior of the DEAE-

cellulose ion exchangers is probably not accessible to larger

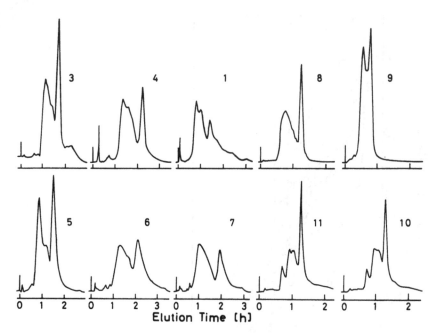

Fig. 7.7 Elution curves of calf serum obtained by linear gradient elution on several commercial ion exchangers. Column size; 1.6 × 15 cm. Initial buffer; 0.05 M Tris-HCl(pH 8.6). Final buffer; 0.05M Tris-HCl(pH 8.6) containing 0.5 M NaCl. Flow rate; 2 mL/min. The slope of the gradient, 1.25 × 10^{-3} M/mL. The numbers in the figure are the same in Fig. 7.3. The number 11 is DEAE Toyopearl 650S. Reproduced from Kato et al. (1982c) by permission of the authors and Elsevier Scientific Publishing Co.

molecular weight substances. Kato et al. (1982c) also measured the elution curves of calf serum by linear gradient elution on various commercial LPIEC and MPIEC columns (Fig. 7.7). It is quite interesting that the resolution and shape of individual peaks show considerable differences although the patterns are similar. A comparison of the separation behavior of proteins was also performed for various preparative MPIEC column by

Larre and Gueguen (1986) and for various HPIEC columns by
Josic et al. (1986). However, we cannot choose the most suitable IEC column on the basis of these results, since the best
chromatographic conditions for a given protein vary with the
properties of an individual IEC column.

Several new types of commercial IEC columns and packings
appear every year. These columns are usually equal or superior to previously available columns.

8
Factors Affecting Separation Behavior

As emphasized previously, the elution mechanism of proteins in ion-exchange chromatography (IEC) is very complicated, since a number of factors are involved in the elution process. For the prediction of operating conditions and/or the design of column dimensions, therefore, it is necessary to know how one particular variable affects the elution (separation) behavior while the others are fixed.

In this chapter, the effects of various factors on elution (separation) behavior are described first for linear gradient elution and then for stepwise elution. In Sec. 8.1, the effect of various operating conditions in linear gradient elution on the elution behavior of a single protein is discussed. Since the effects of the slope of the gradient on the peak position and peak width are very important for optimization and/or scaling up, they are discussed in detail in Sec. 8.2 not only in terms of the rigorous model presented in Sec. 2.2.2 but also in terms of the simple methods given in Sec. 2.2.3.

Section 8.3 deals with the separation mechanism of two components in terms of resolution, defined as the distance between
two adjacent peak maxima divided by the average bandwidth ex-
trapolated to the baseline. Factors affecting resolution are discussed on the basis of the resolution equation for linear gradient
elution, Eq. (2.92), derived in Sec. 2.2.4. In order to test the
validity of this resolution equation, the experimental results in
the literature are compared with the results predicted by the
equation.

In Sec. 8.4, elution behavior in stepwise elution is illustrated
in contrast to that in linear gradient elution. Several additional
elution procedures are described in Sec. 8.5. Other special
operational methods and apparatus are also reviewed briefly in
Sec. 8.5. Finally, in Sec. 8.6, the ranges of applicability of the
model given in Sec. 2.2.2, the simple methods in Sec. 2.2.3,
and the resolution equation in Sec. 2.2.4 are summarized. Some
discussion of the retention behavior is also presented in that
section. The versatility of high-performance IEC (HPIEC), that
is, a high speed of separation and high resolution, is demonstrated with respect to particle size in Secs. 8.1.6 and 8.3.4.

8.1 ELUTION BEHAVIOR IN LINEAR GRADIENT ELUTION

In this section, the effects of various factors on elution behavior are illustrated, based principally on our previous study
(Yamamoto et al., 1983b), so that one can easily determine how
one operating variable affects elution behavior. Experimental
elution curves of a single protein under various experimental
conditions are compared with those calculated by the model presented in Sec. 2.2.2. In order to avoid the contribution of the

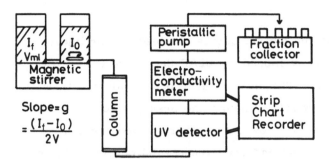

Fig. 8.1 Block diagram of the setup employed for the linear gradient elution experiment in Yamamoto et al. (1983b).

heterogeneity of the sample protein to zone spreading, all the proteins employed are highly purified before use. The linear gradient elutions of proteins were performed using a linear increase in the NaCl concentration in the buffer. The experimental apparatus used is shown schematically in Fig. 8.1. A calculation procedure for the elution curve by the model in Sec. 2.2.2 is found in Appendix D. For further details of the experimental conditions and the calculation procedures, see our original papers (Yamamoto et al., 1983a,b).

The initial concentrations of the sample proteins were first chosen to follow the linear parts of the isotherms. This simplifies the mechanism of elution and the theory since the concentration of the stationary phase can be considered a function of the ionic strength only; that is, $\overline{C} = K[I]C$. Therefore, the theoretical elution curves in Figs. 8.2 through 8.6 were calculated using the relation $K[I] = AI^B + K_{crt}$ [Eq. (3.39)] with the values of A, B, and K_{crt} shown in Table 3.1. The effect of the initial sample concentration is discussed in Sec. 8.1.4.

8.1.1   Flow Rate

Figure 8.2 shows the elution curves obtained by the linear
gradient elution of ovalbumin on DEAE-Sepharose CL-6B with
various flow rates.   The elution curve became wider with an
increase in the flow rate although the peak positions were constant regardless of the flow rate.   There is good agreement
between the theoretical and the experimental curves.   The number of plates N_p was determined according to Eq. (2.75) with
the parameters shown in Tables 3.1, 4.2, and 5.3 and the distribution coefficient at the ionic strength of peak position K_R.
The value of K_R is calculated by Eq. (3.39) with the value of
the ionic strength of the peak position I_R of the experimental
elution curve.   An increase in the flow rate reduces the N_p
value, as is clear from Eq. (2.75).   Detailed examination of the
theoretical elution curves in Fig. 8.2 showed that the maximum
concentration of the theoretical elution curve is directly proportional to the square root of N_p and the variance with respect
to θ, σ_θ^2, is inversely proportional to N_p.   These relations,
which are the same as those in the standard plate theory [see
Eq. (2.38)], were confirmed as valid for symmetrical elution
curves by numerical calculations for a variety of experimental
conditions.   Therefore, once an elution curve is calculated by
our model for a given set of experimental conditions with a
certain value of N_p, the elution curves for the same experimental conditions, except for flow rate, are readily predictable
by use of N_p values corresponding to the respective flow rate
without calculating the entire elution curve for each flow rate.
The number of plates for NaCl N_p' is estimated according to Eq.
(2.76), with the assumptions that $\overline{D}$ for NaCl, $\overline{D}'$, is one-fifth

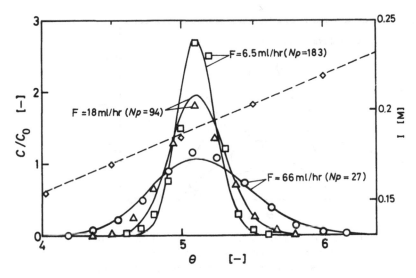

Fig. 8.2 Effect of flow rate in the linear gradient elution of ovalbumin on DEAE-Sepharose CL-6B. Here, θ is the dimensionless time or volume; that is, θ = Ft/V$_0$ = V/V$_0$, where V = elution volume, t = elution time from the start of elution, and F = flow rate. Experimental conditions are as follows; particle diameter d$_p$ = 110 μm; column diameter d$_c$ = 1.5 cm; column length Z = 10 cm; sample (ovalbumin) volume = 6mL; sample concentration C$_0$ = 0.021%; initial ionic strength I$_0$ = 0.11 M; the slope of the gradient g = 5 ×10^{-3} M/mL; pH = 7.9; and temperature = 20°C. The calculated results for the ionic strength for F = 0.11 mL/min (N$_p'$ = 400), F = 0.3 mL/min (N$_p'$ = 430), and for F = 1.1 mL/min (N$_p'$ = 300) were coincident and therefore represented by the single broken line in the figure. (◇) The experimental results for the ionic strength; other symbols represent the experimental results for the protein. (From Yamamoto et al., 1983b.)

the molecular diffusion coefficeint and D$_L$/u is equal to the

particle size of the ion exchanger d$_p$ (for the details of these

relations, see Chaps. 4 and 5). The results calculated for NaCl

(ionic strength I) are represented by a broken line in the figure.

Extensive calculations showed that the linear gradient at the

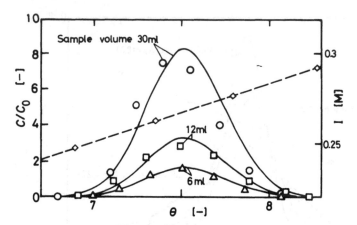

Fig. 8.3 Effect of sample volume in the linear gradient elution of β-lactoglobulin A on DEAE-Sepharose CL-6B. The solid curves are the results calculated for the protein with N_p = 107 and the broken line for the ionic strength with N_p' = 400. ($\circ$, $\triangle$, $\square$) The experimental results for the protein; ($\diamond$) the result for the ionic strength. Experimental conditions were d_p = 110 μm, d_c = 1.5 cm, Z = 10 cm, C_0 = 0.025%, I_0 = 0.11 M, g = 5 × 10^{-3} M/mL, F = 0.3 mL/min, pH 7.9, and 20°C. (From Yamamoto et al., 1983b.)

exit of the column dispersed until θ = 1 + HK' + 0.15 when N_p' is greater than 100, and then it can be approximately expressed by

$$I_{(N_p')} = I_{(0)} + G[\theta - (1 + HK')] \tag{8.1}$$

Furthermore, the effect of N_p' on the elution curves of proteins was found to be negligible when N_p' was greater than 100. The elution curves of proteins calculated with N_p' = 100−400 and those without considering N_p', that is, I is represented by Eq. (2.70) and dI/dθ by Eq. (2.71), were almost the same. The effect of N_p' is important only in stepwise elution using an elution buffer of high ionic strength, as is shown in Sec. 8.4.

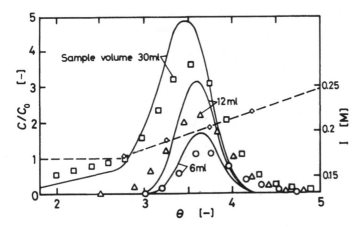

Fig. 8.4 Effect of sample volume in the linear gradient elution of bovine serum albumin on DEAE-Sepharose CL-6B. The solid curves are the results calculated for the protein with N_p = 97 and the broken line is the ionic strength with N'_p = 400. ($\circ$, $\triangle$, $\square$) The experimental results for the protein; ($\diamond$) the result for the ionic strength. Experimental conditions were d_p = 110 μm. d_c = 1.5 cm, Z = 10 cm, C_0 = 0.03%, I_0 = 0.17 M, g = 5 × 10^{-3} M/mL, F = 0.3 mL/min, pH 7.9, and 20°C. (From Yamamoto et al., 1983b.)

8.1.2 Sample Volume

Figure 8.3 shows the elution curves of β-lactoglobulin A on DEAE-Sepharose CL-6B with various sample volumes. The distribution coefficient at the initial ionic strength K_0 was more than 100. An increase in sample volume up to 30 mL did not affect the peak positions and the peak shapes, as shown in the figure. The standard deviations for these curves were almost the same, which implies that the maximum concentration of the elution curve C_{max} is directly proportional to the sample volume. On the other hand, when the value of K_0 was not large (less than 5), the peak position shifted to a smaller value of θ and the peak became asymmetric with an increase in sample volume, as shown in Fig. 8.4. In this case, the N_p values for three elution curves

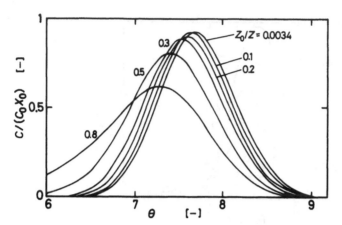

Fig. 8.5 Calculated elution curves for various sample volumes.
The solid curves are the results calculated with N_p = 30. Here,
X_0 is the ratio of the sample volume to the void volume of the
column V_0; X_0 is chosen so that Z_0/Z calculated by Eq. (8.2)
represents the value shown in the figure. The same experimental conditions as in Fig. 8.3 are assumed, that is, a linear
gradient elution of β-lactoglobulin A on DEAE-Sepharose CL-6B.
The calculated line for the ionic strength is not shown, since
it was coincident with that in Fig. 8.3. (From Yamamoto et al.,
1983b.)

were assumed to be the same and were calculated by Eq. (2.75)
with K_R for the elution curve of sample volume = 6 mL. Deviations of the calculated from the experimental results in Fig.
8.4 may be due to the microheterogeneity of bovine serum albumin
(BSA) that has been reported by several investigators (Janatova
et al., 1968; Hartley et al., 1962).

Although these phenomena, that is, the asymmetric elution
curve and the shift of the peak position caused by an increase
in sample volume, are unfavorable, there has not yet been any
criterion for optimum sample volume. Extensive calculations by
the present model revealed that the ratio of the length of the
sample zone initially adsorbed on the column Z_0 to the column

length Z should be smaller than 0.2—0.3, as seen in Fig. 8.5. For practical purposes, Z_0/Z may be expressed by the equation

$$\frac{Z_0}{Z} = \frac{X_0}{1 + HK_0} \qquad (8.2)$$

where X_0 = ratio of sample volume to the void volume of the column.

8.1.3 Gradient Slope

Figure 8.6 shows the effect of the slope of the linear gradient for ovalbumin on DEAE-Sepharose CL-6B. As the slope of the gradient g decreases, the peak position shifted to a larger θ value, that is, a larger elution volume, and the peak width increased. Moreover, the ionic strength at the peak position I_R decreased with the decrease in g. These phenomena are discussed in detail in Sec. 8.2.

8.1.4 Initial Sample Concentration

In the early stages of purification processes and on the preparative scale of fractionation, the protein concentration of the sample C_0 is frequently so high that the protein obeys a nonlinear adsorption isotherm. Figure 8.7 shows the effect of the initial protein concentration in linear gradient elution. As the initial concentration of the sample protein C_0 was increased, the experimental elution curve became asymmetric, the amount of tailing increased, and the peak position shifted to a smaller θ. Similar experimental results are often found in the literature (for example, Kato et al., 1982b). These phenomena are quite similar to that of the isocratic elution of a solute following a nonlinear adsorption isotherm (see Sec. 2.1.2 and Fig. 2.4).

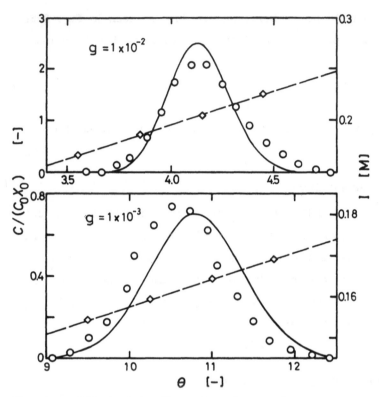

Fig. 8.6 Effect of gradient slope in the linear gradient elution of ovalbumin on DEAE-Sepharose CL-6B. The solid curves are the results calculated for the protein with N_p = 96 for g = 0.01 M/mL and N_p = 64 for g = 0.001 M/mL, and the broken lines for the ionic strength with N_p' = 400. ($\diamond$, o) The experimental results for the ionic strength and the protein, respectively. Experimental conditions were d_p = 110 μm, d_c = 1.5 cm, Z = 10 cm, C_0 = 0.2%, sample volume 6 mL for g = 0.01 M/mL and 20 mL for g = 0.001 M/mL, I_0 = 0.11 M, F = 18 mL/h, pH 7.9, and 20°C. Note that the scales of the horizontal and vertical axes are different for the two elution curves. (From Yamamoto et al., 1983b.)

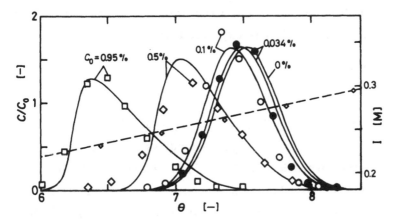

Fig. 8.7 Effect of initial concentration of the sample in the linear gradient elution of β-lactoglobulin A on DEAE-Sepharose CL-6B. The solid curves are the results calculated for the protein with N_p = 107 and the broken line for the ionic strength with N'_p = 400. (◊) The experimental results for the ionic strength; the other symbols represent the results for the protein. Experimental conditions were d_p = 110 μm, d_c = 1.5, Z = 10 cm, sample volume 6 mL, I_0 = 0.11 M, g = 5 × 10^{-3} M/mL, F = 0.3 mL/min, pH 7.9, and 20°C. (From Yamamoto et al., 1983b.)

The present model is also applicable when the adsorption iso-therms are available as a function of ionic strength. It is time consuming to measure a number of adsorption isotherms over a wide range of ionic strengths, however. Therefore, the ad-sorption isotherms of β-lactoglobulin A on DEAE-Sepharose CL-6B were re-examined. Approximately linear relation were observed for the isotherms at various ionic strengths between the reciprocal of the concentration of the protein in the ion ex-changer $1/\overline{C}$ and that in the solution $1/C$ with different slopes and almost the same intercepts. This indicates that adsorption isotherms follow the Langmuir equation, and the maximum amount of the protein adsorbed Q_{max} is constant regardless of ionic

strength under the experimental conditions examined here. The
relation between $\overline{C}$ and C is expressed as

$$\overline{C} = \frac{qQ_{max}C}{Q_{max} + qC} \qquad\qquad (8.3)$$

where $q = K[I] = AI^B + K_{crt}$ and $Q_{max} = 5.71\%$. Although the
assumption of the constant Q_{max} regardless of I seems contradictory to the discussion in Chap. 3, it is employed here for
simplification of the calculation procedure and for qualitative
investigation of elution behavior. By using Eq. (8.3) and the
same N_p value employed in Fig. 8.3, the theoretical elution
curves in Fig. 8.7 were calculated. It is seen from Fig. 8.7 that
the calculated theoretical curves predicted the experimental
curves fairly well. Although the initial concentration of the
sample proteins, $C_0 = 0.034\%$, is greater than the range of the
quasi-linear part of the isotherms, the theoretical elution curve
for $C_0 = 0.034\%$ is almost symmetrical and its shape and position are quite similar to those of the theoretical curve for the
limiting case $C_0 = 0\%$. This suggests that the range of the initial sample concentration where the assumption of linear isotherms would not lead to serious error is relatively wider than
that determined from the isotherms for initial ionic strength.
This is favorable for practical purposes and may be because the
initial quasilinear part of the isotherm becomes wider with the
increase in ionic strength during a linear gradient elution, as
mentioned in Chap. 3 and as previously reported (Morris and
Morris, 1964). The deviations of theoretical elution curves from
experimental curves in Fig. 8.7 may be due to the assumptions
of constant Q_{max} and of the same N_p value for different C_0.
If the data of adsorption isotherms are measured over a wide

range of ionic strength at small successive intervals, agreement between the theoretical and experimental elution curves would be better.

8.1.5 Temperature

Ion-exchange chromatography is often operated at subambient temperature for the separation of unstable enzymes and proteins. Temperature may affect both the parameters in the height equivalent to a theoretical plate (HETP) or N_p equation and the adsorption equilibrium between proteins and ion exchangers. As described in Sec. 2.1.6, HETP (peak width) increases with a decrease in temperature. Therefore, a decrease in temperature is not advantageous from the point of view of column efficiency.

Although the extent of temperature dependence of adsorption equilibria is needed to anticipate how the peak retention time varies with temperature, to our knowledge there are few available experimental data or theoretical studies. James and Stanworth (1964) reported that a change in temperature between 0 and 50°C did not affect the adsorption capacity of human serum proteins on DEAE-cellulose. Frolik et al. (1982) observed a slight increase in the retention time of proteins and a significant decrease in the protein recovery on high-performance cation-exchange columns when the temperature was decreased from 55 to 25°C. Figure 8.8 shows the elution behavior of a mixture of three proteins (ovalbumin and β-lactoglobulins A and B) on a DEAE ion-exchange gel column (Yamamoto et al., unpublished data). It is seen that both the resolution of β-lactoglobulins A and B and the maximum peak height of the three proteins decrease with a decrease in temperature. A slight decrease in peak retention time is also observed with a decrease in tempera-

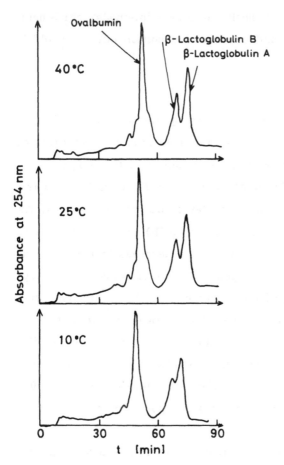

Fig. 8.8 Effect of temperature on the elution behavior of proteins in linear gradient elution. The linear gradient elution was
performed by a linear increase in NaCl concentration at pH 8.5
with a DEAE-Toyopearl 650S MPIEC column (d_p = 40 μm, d_c =
1.6 cm, Z = 15 cm). g = 9 × 10^{-4} M/mL, the initial buffer, 4
mM Tris HCl (pH 8.5) containing 0.03 M NaCl, F = 3.0 mL/min.
Sample, 1 mL 0.5% ovalbumin + 0.5% β-lactoglobulins A and B.

ture. The application of IEC to protein separation at subzero temperatures was reproted previously (Douzou and Balny, 1978).

Care should also be taken with the change in the pH of the elution buffer with the temperature, as pointed out by Peterson (1970) and Scopes (1982). For example, the pH of a buffer solution prepared with Tris (hydroxymethylaminomethane), and HCl increases an average of 0.03 pH units per degree Celsius as the temperature of the solution decreases from 25 to 5°C [Sigma Tech. Bulletin. No. 106B(11-78)].

In addition, as the temperature is lowered, the pressure drop increases owing to an increase in the viscosity of the elution buffer. For example, the viscosity of water is 1.57 cP at 4°C but it is 0.89 cP at 25°C. This indicates that the pressure drop at 4°C is 1.8 times higher than that at 25°C, according to the Kozeny-Carman equation (7.1).

It is therefore not advantageous to operate IEC at subambient temperatures unless the enzymes or proteins to be separated are extremely unstable. In certain cases, a rapid separation at room temperature may be superior to separation at subambient temperatures even for the separation of unstable substances.

8.1.6 Particle Diameter

If the ion-exchange capacity and the gel matrix are the same, the change in particle diameter d_p only contributes to zone spreading. In Sec. 2.1.6, we described how the column efficiency (the total number of plates N_p = Z/HETP) increases drastically with decreasing d_p. Therefore, when d_p is reduced, although the peak position is not varied the peak is narrower. This expectation is shown to be valid by the experimental results shown in Sec. 8.3.4. It is again stressed that this is why high-performance ion-exchange chromatography gives a high separation efficiency. However, at the same time, the pressure drop per unit length also increases with the inverse of the square of d_p according to the Kozeny-Carman equation (7.1). Therefore, if d_p is reduced from 120 to 40 µm, the pressure force must be increased by a factor of 9 to obtain the same flow rate. HPIEC packings and columns must be resistant to such high-pressure forces. The effect of d_p on separation efficiency is described in detail in Sec. 8.3.4.

8.1.7 Column Dimension

The model presented in Sec. 2.2.2 predicts that the dimensionless peak position θ_R is constant and the dimensionless peak width σ_θ is inversely proportional to the column length when the dimensionless slope of the gradient G (= gV_0) and linear velocity u are fixed. The situation is much more complicated if G is not constant. This is discussed in detail in Sec. 8.3.

If the flow rate is varied in proportion to the cross-sectional area of the column, the same elution behavior is observed. However, when soft compressible gels are employed, the pressure drop per unit length as well as the HETP depends on the column diameter and length (Janson and Hedman, 1982). The problems in scaling up IEC columns are discussed in Chap. 9.

8.1.8 Recovery

The recovery of proteins or enzyme activities from IEC columns is usually high, more than 80—90% (Kopaciewicz and Regnier, 1983; Kato et al., 1982a; 1983b; 1985b; Yamamoto et al., 1987b). However, low protein recoveries are observed in certain cases

(Kopaciewicz and Regnier, 1983; Kato et al., 1985b). Kato et al. (1985b) investigated various factors affecting protein recovery, such as the slope of the gradient, the flow rate, and the initial pH, using proteins of different isoelectric points pI. They found that the protein recovery was seldom affected by the flow rate and increased slightly with an increase in the slope of the gradient. The most important factor influencing the protein recovery, however, was the relation between the pH and pI of proteins. Kato et al. (1985b) showed that acidic proteins, such as thyroglobulin, ferritin, bovine serum albumin, and trypsin inhibitor, and neutral proteins, such as hemoglobin and myoglobin, were not recovered from a high-performance strong cation-exchange column at pH 4.0. They stated that the separation of acidic and neutral proteins on strong cation-exchange columns at low pH is not generally recommended. Kopaciewicz and Regnier (1983) also reported similar results. They stated that this low protein recovery was caused by a change in the protein structure. In addition, they have pointed out that the recovery also varies with the type of salt. Frolik et al. (1982) found that the recovery of carbonic anhydrase from strong cation-exchange columns increased with temperature in the temperature range 25−65°C. Therefore, we should investigate these chromatographic conditions when the recovery of a desired protein is low.

Most of the variables discussed in this section (Sec. 8.1) are examined again in the following section in terms of the resolution of two peaks.

8.2 PREDICTION OF PEAK POSITION AND PEAK WIDTH IN LINEAR GRADIENT ELUTION BY SIMPLIFIED METHODS

As mentioned in Sec. 8.1.3, the peak width becomes wider and the ionic strength (salt concentration) at the peak position I_R decreases with a decrease in the slope of the linear gradient. Occasionally, the method called rechromatography is employed in order to check the homogeneity of the sample or to perform further purification. Although the pH and the ion exchangers are the same as those employed for the first chromatography, the column volume and the slope of the gradient are somewhat different from those used for the first chromatography. This gives rise to a shift in the peak position and I_R, which is confusing in testing the homogeneity of the fractionated protein and may lead to misunderstanding of the results. Similar problems are encountered in scaling up of the separation, since the column dimension and the flow rate as well as the slope of the gradient must be changed because of the mechanical stability of the packing materials, as described in Chaps. 7 and 9. A method of predicting peak position and peak width in linear gradient elution is thus needed.

As demonstrated in the previous section, the peak position and the peak width in linear gradient elution can be predicted by numerical calculations of the whole elution curve by the model presented in Sec. 2.2.2. However, since this numerical calculation requires a high-speed, large-memory computer, any simpler methods would be helpful and useful from a practical point of view. In Sec. 2.2.3, a simple graphic method for predicting I_R and an asymptotic solution for the peak width are presented. We demonstrate the validity and applicability of this

method and equation by comparing the values obtained by them
with the calculated results from the model given in Sec. 2.2.2
and with experimental results.

8.2.1 Peak Position

In Sec. 2.2.3, the following equations are derived for the relations between the ionic strength (salt concentration) at the
peak position I_R and the slope of the gradient g (see Fig. 2.17):

$$h(I_R) - h(I_0) = GH \tag{2.83}$$

$$h(I) = \int_{I_a}^{I} \frac{dI}{K(I) - K'} \tag{2.82}$$

where I_0 = initial ionic strength (salt concentration)

$H = (V_t - V_0)/V_0$

$G = gV_0$ = slope of the gradient normalized with respect to
the column void volume V_0

V_t = total column volume

I_a = arbitrary I value less than I_0

$K(I)$ = distribution coefficient of a protein, which depends on I

K' = distribution coefficient of a salt, which is usually constant

$GH = g(V_t - V_0)$ is also regarded as the slope of the gradient
normalized with respect to the column gel volume.

The solid curves in Fig. 8.9 show the results calculated by
Eq. (2.82) where linear adsorption isotherms are assumed; that
is $K = K[I] = AI^B + K_{crt}$. The values of A, B, K_{crt}, and K'
shown in Table 3.1 were employed in these calculations. The
results calculated from the model given in Sec. 2.2.2 agreed
well with those from Eq. (2.82). Moreover, the experimental
results, in which the sample concentration was low enough to

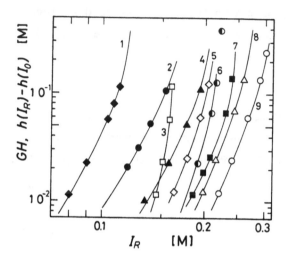

Fig. 8.9 Relations between GH and I_R. The solid curves are the results calculated by Eq. (2.82) in Sec. 2.2.3 with the data shown in Table 3.1. (1) Ovalbumin on DEAE-Bio-Gel A; (2) ovalbumin on DEAE-Toyopearl 650; (3) ovalbumin on DEAE-Sephadex A-25; (4) bovine serum albumin on DEAE-Sepharose CL-6B (pH 6.1); (5) ovalbumin on DEAE-Sepharose CL-6B; (6) bovine serum albumin on DEAE-Sepharose CL-6B; (7) ovalbumin on DEAE-Sephadex A-50; (8) β-lactoglobulin B on DEAE-Sepharose CL-6B; (9) β-lactoglobulin A on DEAE-Sepharose CL-6B. Experimental conditions were d_c = 1.5 cm Z = 10 cm; sample volume, 6—20 mL; C_0 = 0.02—0.03%; I_0 = 0.11 M (for ovalbumin on DEAE-Bio-Gel A and Toyopearl, I_0 = 0.03); F = 0.3 mL/min; pH 7.9, and 20°C, unless otherwise indicated. (From Yamamoto et al., 1983b).

follow the initial quasi-linear isotherm and the sample volume was chosen to be adsorbed as a thin layer at the top of the column, showed good agreement with the solid curves shown in Fig. 8.9. As the gradient becomes steep, I_R increases and finally becomes constant when the K_R value becomes equal to K'. This is shown not only by theoretical results but also by the experimental results of Fig. 8.9. The value of I at which K = K' is very important since it reflects the intrinsic adsorption properties of proteins. If the protein zone is subjected to this I value (hereafter abbreviated I') in the linear gradient elution, it moves with the same velocity of the elution buffer and a quasi-steady state, explained in Sec. 2.2.3, is attained. I' is also an important parameter in stepwise elution and is discussed in Sec. 8.4.

In a DEAE-Sephadex A-25 ion exchanger, the distribution coefficient of NaCl between the solution and the ion exchanger K' was dependent on the ionic strength of the solution, as reported by Kirkegaard (1976) and by Johnson and Bock (1974). Therefore, the K' value for DEAE-Sephadex A-25 was not readily determined. However, the calculated results with the assumption that K' = 0.5 is suitable, for practical purposes, for the prediction of I_R.

When K(I) is unknown, linear gradient elution experiments with different slopes of the gradient are carried out with a given small IEC column. GH is then plotted against the measured I_R value. As shown in Fig. 8.10, the GH versus I_R curve thus obtained is not dependent on the column dimension, the flow rate, or the particle diameter. The data of Kato et al. (1982b), in which I_R was measured as a function of the gradient time t_G with HPIEC, were also found to be expressed by a single relation between GH and I_R. Therefore, once a chart like Fig. 8.9 or 8.10 is prepared by numerical integration of Eq. (2.82) with experimentally measured K(I) or by linear gradient elution experiments with a small column, we can determine I_R with four parameters: initial ionic strength I_0, the slope of the gradient g, the column void volume V_0, and the total column volume V_t. For calculation of the peak position (retention time) t_R, the distribution coefficient for a salt K' is

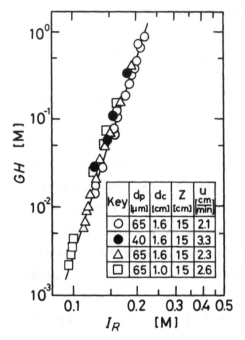

Fig. 8.10 Relations between GH and I_R for various experimental conditions. Sample, 0.1% ovalbumin 5 mL. Column, DEAE-Toyopearl 650S and 650M. Initial condition, 14 mM Tris-HCl buffer (pH 7.7) containing 0.03 M NaCl (ionic stength = 0.04). The linear gradient elution was performed by increasing NaCl concentration in the buffer at 25°C. The measured I_R (= NaCl concentration + 0.01) values were plotted against GH = $g(V_t - V_0)$. (From Yamamoto et al., 1987d.) Reproduced by permission of the American Institute of Chemical Engineers.

also required. We give one example for determining the I_R of β-lactoglobulin A on DEAE-Sepharose CL-6B shown in Fig. 8.9.

Let us employ the following values: $I_0 = 0.21$, $V_t = 30$ mL, $V_0 = V_t \times 0.4 = 12$ mL, g = 3×10^{-3} M/mL, and K' = 0.9.

1. Read the value of the vertical axis at $I_R = I_0$; answer, 7×10^{-3}.
2. Let add this value from GH = $g(V_t - V_0)$; answer, $3 \times 10^{-3} \times (30 - 12) + 7 \times 10^{-3} = 0.061$.

3. Read out the value of the horizontal axis corresponding to the value of the vertical axis equal to 0.061; answer, I_R = 0.274.

4. Calculate θ_R by Eq. (2.84); answer, $\theta_R = (I_R - I_0)/G + 1 + HK' = (0.274 - 0.21)/(3 \times 10^{-3} \times 12) + 1 + (30 - 12)/12 \times 0.9 = 4.13$.

5. Obtain the elution volume V_e by multiplying V_0 by θ_R; answer, $V_e = V_0 \theta_R = 12 \times 4.13 = 49.6$ mL.

Therefore, the time when the peak is eluted from the column t_R can be calculated by $t_R = V_e/$flow rate (mL/min). It should be mentioned that procedure 1 can be skipped under the usual chromatographic conditions since the value of the vertical axis at $I_R = I_0$ is negligibly small.

The calculated values of θ_R according to this procedure are shown in Fig. 8.11 for β-lactoglobulins A and B on DEAE-Sepharose CL-6B. It can be seen that θ_R increases with a decrease in the slope of the gradient.

Since the left-hand side of Eq. (2.83) is numerically integrated, the explicit relation between I_R and the slope of the gradient or the ionic strength dependence of the distribution coefficient is difficult to anticipate. Therefore, we assume that the ionic stength-dependent distribution coefficient K(I) can be approximated to

$$K[I] = A'I^{-B'} + K' \qquad (8.4)$$

where A' and B' = constants and K' = distribution coefficient of a salt. Actually, this equation is a good approximation of the experimental results shown in Chap. 3 when K_{crt} is close to K'. When Eq. (8.4) is inserted into Eq. (2.81), the following analytical solution is obtained:

$$GH = \frac{I_R^{(B'+1)} - I_0^{(B'+1)}}{A'(B' + 1)} \qquad (8.5)$$

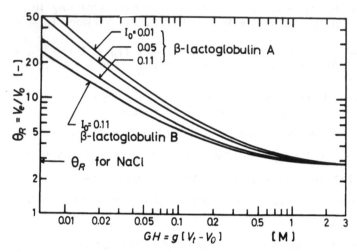

Fig. 8.11   Relations between θ_R and GH. θ_R was calculated according to the procedure shown in the text with the data shown in Fig. 8.9. I_0 is the initial ionic strength[M].

Since $I_0^{(B'+1)}$ is negligibly small compared with $I_R^{(B'+1)}$ under the usual chromatographic conditions, Eq. (8.5) is simplified to

$$GH = \frac{I_R^{(B'+1)}}{A'(B'+1)} \qquad\qquad (8.6)$$

It follows from this equation that I_R has a power-law dependence on GH.   As discussed in Chap. 3, A' and B' imply the ion-exchange capacity and the number of charges involved in the adsorption process, respectively.   Therefore, as the number of the net charge of proteins is increased (this is caused by an increase in pH for anion exchangers and by a decrease in pH for cation exchangers), the dependence of I_R on GH becomes strong.   In other words, it is expected that the curve of GH versus I_R shifts to larger I_R values and its slope becomes steep with the increase in the net protein charge (see Fig. 10.2).

These expectations are shown to be valid by the experimental
results in Fig. 8.9, in which the curve for BSA on DEAE-Sepharose CL-6B at pH = 7.9 locates to higher I_R values and is
steeper than that at pH = 6.1. Since the increase in pH from
6.1 to 7.9 causes an increase in the net charge BSA (the pI
of BSA is around 4.8), this experimental result verifies Eq.
(8.6). Therefore, we can expect this plot of GH versus I_R as
a rapid method both for characterizing the nature of proteins
and for surveying the optimum chromatographic conditions, if
the plot is prepared for various pH values. This is discussed
in detail in Sec. 8.6.

As shown in Sec. 2.2.3, the GH versus I_R curve is also
employed for the determination of K(I) on the basis of Eq.
(2.85):

$$\frac{d(GH)}{dI} = \frac{1}{K(I) - K'} \tag{2.85}$$

As shown in Fig. 8.12, the K(I) values obtained from differentiation of the GH versus I_R curve in Fig. 8.10 are in good
agreement with those by isocratic elution (pulse response) experiments and by batch experiments (see Sec. 3.2). The
values of K(I) is needed to design the stepwise elution of proteins on IEC columns (see Secs. 8.4 and 10.2). Actually, it is
time consuming and laborious to determine K(I) by the batch
experiments shown in Sec. 3.3 when the K value is 1 ~ 10 and/
or the contaminant proteins are included in the sample. It is
therefore recommended that the GH versus I_R curve be prepared with a given small IEC column. The curve obtained is
useful for characterizing the nature of proteins, for surveying
the optimum chromatographic conditions, and for determining
K(I).

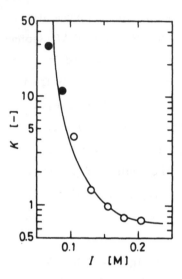

Fig. 8.12 Relation between the distribution coefficient K of ovalbumin and the ionic strength I. Open and closed circles are the data obtained by isocratic elution and by batch experiments (see Sec. 3.2), respectively. The solid curve was obtained from differentiation of the curve in Fig. 8.10. (From Yamamoto et al., 1987d.) Reproduced by permission of the American Institute of Chemical Engineers.

8.2.2 Peak Width

The prediction of elution curves for linear gradient elution by the model presented in Sec. 2.2.2 was successful, as shown in the previous section, especially for symmetrical elution curves. Furthermore, it was found that the effect of the number of plates for salts N_p' was negligible and that the variance of the elution curve of proteins σ_θ^2 was inversely proportional to the number of plates for proteins N_p. Since I_R can be easily obtained by the method described in Sec. 8.2.1, we are now able to determine the N_p value for a given set of experimental conditions by insertion in Eq. (2.75) of the parameters obtained by the moment method and the operating conditions. Therefore, if the relations among σ_θ^2, N_p, and the operating conditions, such as G, are known, the peak width in linear gradient elution can be predicted.

In Sec. 2.2.2, we derived the following asymptotic solution from a quasi-steady-state model for the prediction of the peak width in the linear gradient elution:

$$\frac{\sigma_\theta^2 N_p}{(1 + HK_R)^2} = -\frac{1}{2} \frac{1 + HK_R}{GH(1 + HK')(dK/dI)_{I=I_R}} \tag{2.86}$$

On the basis of Eq. (2.86), the correlation shown by Fig. 8.13 was attempted between the theoretically calculated results from the model given in Sec. 2.2.2 and the experimental results. Since asymmetric elution curves are unfavorable for good separation, we are concerned only with the experimental conditions in which symmetrical elution curves are obtained; that is, $Z_0/Z < 0.1$ and $\overline{C} = K[I]C$. The results calculated by the model given in Sec. 2.2.2 for various proteins and ion exchangers over a wide range of experimental conditions fell in the region between the two solid curves shown in Fig. 8.13. These two curves approach the approximate solution by the quasi-steady-state model with a decrease in the value of the abscissa. In other words, the quasi-steady state is attained only for the limiting conditions of the steep slope of the gradient, long column length, and a sharp decrease in K with the increase in I. These conditions, however, are not always fulfilled under the usual experimental conditions. The experimental results for various proteins and ion exchangers are scattered around the two curves from the model shown in Sec. 2.2.2. This implies the applicability and utility of this correlation.

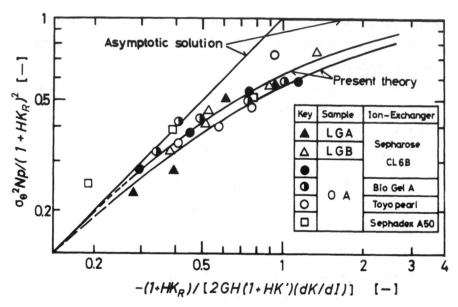

Fig. 8.13 Relations between $\sigma_\theta^2 N_p/(1 + HK_R)^2$ and $-(1 + HK_R)/$ $[2GH(1 + HK')dK/dI]$. The σ_θ^2 of the numerically calculated elution curves by the model described in Sec. 2.2.2 was determined by measuring the peak width at $C/C_{max} = 0.3678$, where C_{max} is the maximum concentration of the elution curve (the peak width measured by this procedure is $2\sqrt{2}\sigma_\theta$; see Fig. 2.1). The results using the model in Sec. 2.2.2 were correlated between the two curves shown in the figure. The straight line in the figure is the asymptotic solution represented by Eq. (2.86) in Sec. 2.2.3. The horizontal straight line at $\sigma_\theta^2 N_p/(1 + HK_R)^2$ = 1.0 is another asymptotic solution given by Eq. (2.87) in Sec. 2.2.3. The ion exchangers have a DEAE group. (From Yamamoto et al., 1983b.)

Experimental elution curves for DEAE-Sephadex A-25 were highly asymmetric and had a long tail. Therefore, these results are not shown in the fiugre. Asymmetric elution curves with a long tail may be ascribed to a diffusion coefficient in gel that is much lower than the molecular diffusion coefficient. This low diffusion coefficient was reported for glucoamylase on CM-Sephadex

C-25 by Adachi et al. (1978) (See Chap. 3). Knight et al.
(1963) also reported a low diffusion velocity in CM-Sephadex
C-25 compared with other ion exchangers. It may be concluded
that DEAE-Sephadex A-25 ion exchangers are not suitable for
the proteins employed in this study. This ion exchanger should
be employed for the separation of substances whose molecular
weight is less than 10,000 or greater than 400,000, according
to the manufacturer (Pharmacia Fine Chemicals, 1987).

The two curves obtained with this model will probably approach unity with an increase in the value of the abscissa,
which means that the zone sharpening effect of the gradient
diminishes and the situation becomes similar to an ordinary isocratic elution; that is,

$$\frac{\sigma_\theta^2 N_p}{(1 + HK_R)^2} = 1.0 \tag{2.87}$$

The moment method is not applicable to DEAE-Sephadex A-25
and A-50 ion exchangers, since the distribution coefficient decreased and approached zero with an increase in ionic strength.
However, for DEAE-Sephadex A-50, the N_p value in linear
gradient elution can be approximately determined by the following procedure with the aid of the correlations shown in Fig.
8.13. Equation (2.75) in Sec. 2.2.2 can be rearranged as

$$N_p = \frac{Z}{\alpha + \beta u H K_R^2 / (1 + HK_R)^2} \tag{8.7}$$

This equation contains two unknown constants α and β. Therefore, if at least two experimental elution curves of different K_R

are obtained, then the corresponding N_p values can be determined by the correlations in Fig. 8.13 using experimental conditions and experimentally measured σ_θ^2. These two N_p values with different K_R values can give the α and β in Eq. (8.7). We measured the peak width at 0.368 times the peak height for the determination of σ_θ^2 (see Fig. 2.1) and the I_R value in order to determine the K_R value. By using these experimental values from the two experimental elution curves of ovalbumin on DEAE-Sephadex A-50, the constants α and β are determined as 0.05 and 136, respectively. The N_p values for the experimental points of DEAE-Sephadex A-50 in Fig. 8.13 were determined by this procedure.

8.3 RESOLUTION IN LINEAR GRADIENT ELUTION

The resolution R_S defined by Eq. (2.51) is useful as a measure of the separation efficiency of two adjacent peaks. As described so far, the R_S of given proteins on a particular ion-exchange column is a complicated function of the column and operating variables:

$$R_S = f(u, d_p, d_c, Z, g, \text{temperature, sample loading}) \quad (8.8)$$

where u = linear mobile-phase velocity

d_p = particle diameter

d_c = column diameter

Z = column length

g = slope of the gradient

It is desirable that the R_S is related to these variables. The R_S can be obtained from the elution curves calculated by the model given in Sec. 2.2.2 when adsorption isotherms as a func-

tion of ionic strength and HETP at high ionic strengths as a
function of flow rate are known.   However, the calculation requires much computation time.

On the other hand, in Sec. 8.2, the applicability of simple
methods for predicting the peak width and position in the linear
gradient elution given in Sec. 2.2.3 was confirmed.   Although
these two simple methods are restricted to a sample protein
solution of low concentration, the results calculated by these
two methods will serve as a basis for predicting the elution behavior of a sample of high concentration and large volume.   The
calculation scheme of R_S by the simple methods is shown in Fig.
8.14.

Unfortunately, no simple analytical relations between R_S and
operating (or column) variables are found in this calculation.
Such relations, if any, would be helpful for optimizing the
separation, especially when the adsorption isotherms and HETP
are unknown.   In Sec. 2.2.4, several interesting relations between R_S and operating conditions are extracted from an R_S equation, Eq. (2.92), which is derived with some simplified assumptions.   These relations are verified on the basis of the R_S values
calculated for β-lactoglobulins A and B on DEAE-Sepharose CL-6B
by the scheme shown in Fig. 8.14 with the parameters shown in
Tables 3.1, 4.2, and 5.3.   The experimental results in the literature are also examined in light of their prediction on the basis
of Eq. (2.92).

8.3.1 Flow Rate

Figure 8.15 shows the relation between R_S^{-2} calculated by the
scheme shown in Fig. 8.14 and the linear mobile-phase velocity
u.   There is a linear relation between R_S^{-2} and u, which is

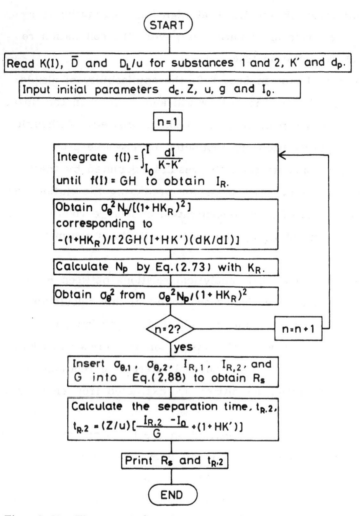

Fig. 8.14 Flowchart for the calculation of R_S in linear gradient elution by the simplified method.

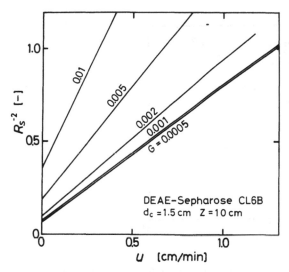

Fig. 8.15   Calculated relations between R_s^{-2} and linear mobile-phase velocity u.   The R_s values for β-lactoglobulins A and B on a DEAE-Sepharose CL-6B column are calculated by the scheme shown in Fig. 8.14 with the values shown in Tables 3.1, 4.2, and 5.3. $G = gV_0$ [M].

predicted by Eq. (2.92).   This relation implies that R_s increases with a decrease in flow rate.   However, since R_s^{-2} is not directly proportional to u, the decreasing flow rate is not always advantageous for increasing R_s, as already pointed out by Vanecek and Regnier (1980).   Furthermore, the separation time increases in inverse proportion to the flow rate.

The experimental results in the literature (Kato et al., 1983a;b) are also expressed by this relation, as shown in Figs. 8.16 and 8.17.   Therefore, if R_s values are measured for various flow rates at a given steepness of the gradient, then the plot of R_s^{-2} against a flow rate will be helpful for predicting the flow rate dependence of R_s.   As described in Sec. 8.3.9,

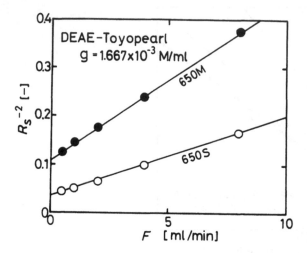

Fig. 8.16 Relations between R_s^{-2} and volumetric flow rate F
for the experimental results of Kato et al. (1983). The particle
diameter of DEAE-Toyopearl 650M and 650S ranges between 44
and 88 μm, and between 25 and 44 μm, respectively. R_s was
measured for ovalbumin and trypsin inhibitor. (Experimental
data are taken from Kato et al., 1983a.)

when the flow rate is high, the simple relation holds that the
R_s is proportional to the inverse of the square root of u.

8.3.2 Column Length

In isocratic elution chromatography, it is expected that R_s is
proportional to the square root of column length Z, as shown in
Eq. (2.55). Therefore, we can increase R_s by increasing Z.
This is because both the moving velocity and the HETP of the
sample in the column are constant during elution (Fig. 8.18).
Therefore, the distance between the two samples increases as
they move down the column. However, in linear gradient elution,
this simple relation for R_s does not hold since the moving velocity of the sample varies with time, as described in Sec. 2.2.1
(see also Fig. 8.18).

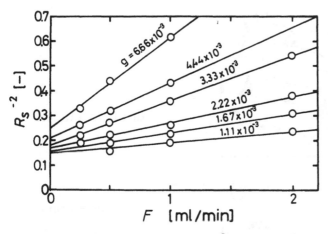

Fig. 8.17 Relations between R_s^{-2} and volumetric flow rate F for the experimental results of Kato et al. (1982b). The HPIEC column (d_c = 0.6 cm, Z = 15 cm) employed is TSK gel IEX-545 DEAE SIL (which is now commercially available as TSK gel DEAE 3SW). The R_s was measured for the two components contained in the commercial ovalbumin. (Experimental data are taken from Kato et al., 1982b.)

Figure 8.19 shows the effect of column length Z on R_s at a fixed slope of the gradient and the same flow rate. R_s increases with Z until a certain Z and then becomes constant. This constant R_s is one of the findings extracted from Eq. (2.92), as shown in Sec. 2.2.4. It is further noted from the figure that as the slope of the gradient g becomes steep, the Z above which R_s becomes constant decreases. This is explained as follows. As illustrated schematically in Fig. 8.18, after a certain time the moving velocities of the two protein zones become equal to that of the elution buffer. The distance between the two zones then no longer varies. This is the quasi-steady state with which the R_s equation (2.92) is derived and the constant R_s is thus obtained there. The quasi-steady state is attained

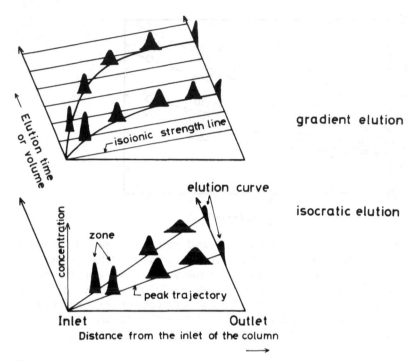

Fig. 8.18 Schematic drawing of the movement of the zones in linear gradient and isocration elution methods. (From Yamamoto et al., 1987d.) Reproduced by permission of the American Institute of Chemical Engineers.

more easily if the slope of the gradient is steep, the column length is long, or the ionic strength dependence of the distribution coefficient is strong. Therefore, R_s for the steep g soon becomes independent of the column length Z but that for a shallow g gradually increases with Z. Figure 8.20 shows the experimental results of the effect of the column length Z on R_s at a fixed slope of the gradient. It is seen that R_s is seldom dependent on Z. Similar experimental results were obtained by Kato et al. (1983a) and by Ekstrom and Jacobson (1984). On the other hand, Vanecek and Regnier (1980) have shown that R_s increases weakly with Z. The column length may not be

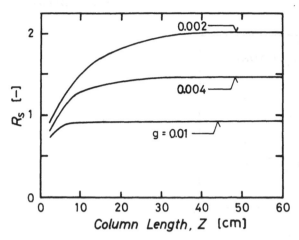

Fig. 8.19 Effect of column length Z on R_S at constant slope g of the gradient. The R_S values for β-lactoglobulins A and B on a DEAE-Sepharose CL-6B column are calculated by the scheme shown in Fig. 8.14 with the values shown in Tables 3.1, 4.2, and 5.3. The values of the parameters employed for the calculation are the same as in Fig. 8.15, except the column length. g[M/mL].

long enough or the slope of the gradient not steep enough to attain the quasi-steady state in their study.

In order to increase R_S with the column length Z, it is necessary to decrease g. The second finding on the effect of column length from Eq. (2.92) is that if g is decreased so that the $G = gV_0$ is constant, R_S increases with the square root of Z. This is shown theoretically in Fig. 8.21 and experimentally in Fig. 8.22. However, it should be borne in mind that both the separation time and the pressure drop increases in proportion to Z.

8.3.3 Gradient Slope

The slope of the gradient is one of the most important operating variables since the separation can be readily optimized by an

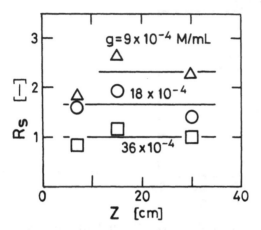

Fig. 8.20 Effect of column length Z on R_S at constant slope of the gradient g. The R_S values for β-lactoglobulins A and B were measured on DEAE-Toyopearl 650S MPIEC columns (particle diameter d_p = 40 μm) at pH 5.6 (10 mM acetate buffer) and 25°C. The linear gradient elution was performed by a linear increase in NaCl concentration from 0.03 M in the buffer solutions. Column diameter d_c = 1.6 cm; linear mobile-phase velocity u = 1.8 cm/min. (From Yamamoto et al., 1987c.) Reproduced by permission of Elsevier Scientific Publishing.

appropriate choice of this variable without any change in column length or in flow rate. However, since the slope of the gradient affects both the peak width and the peak position, as discussed in Sec. 8.2, it is not easy to predict a change in R_S with the slope of the gradient.

It follows from Eq. (2.92) that R_S increases with a decrease in the inverse of the square root of the slope of the gradient $g^{-1/2}$. Figure 8.23 shows the relations between R_S and $g^{-1/2}$ from the calculated results. At low $g^{-1/2}$ values, R_S increases linearly. However, as $g^{-1/2}$ increases, the slope of the R_S versus $g^{-1/2}$ curve gradually decreases. The reason for this decrease in the slope is similar to that for the constant R_S in

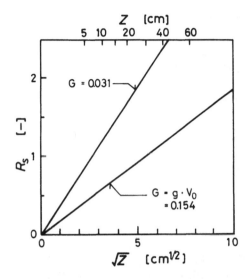

Fig. 8.21   Effect of column length Z on R_S at constant $G = gV_0$.
The R_S values for β-lactoglobulins A and B on a DEAE-Sepharose CL-6B column were calculated by the scheme shown in Fig.
8.14 with the values shown in Tables 3.1, 4.2, and 5.3.   The
values of the parameters employed for the calculation are the
same in Fig. 8.15, except the column length.   G is in M.

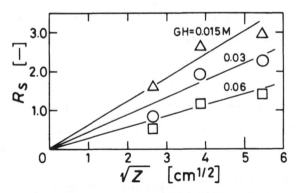

Fig. 8.22   Effect of column length Z on R_S at constant GH.
The experimental conditions are the same as those in Fig. 8.20.
(From Yamamoto et al., 1987c.)

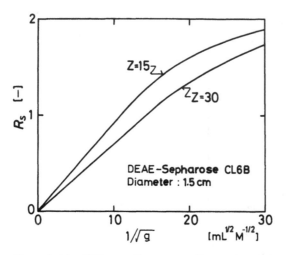

Fig. 8.23 Effect of the gradient slope on R_s. The R_s values for β-lactoglobulins A and B on a DEAE-Sepharose CL-6B column were calculated by the scheme shown in Fig. 8.14 with the values shown in Tables 3.1, 4.2, and 5.3. Z is the column length[cm].

Sec. 8.3.2. That is, a quasi-steady-state approximation gradually becomes invalid with a decrease in the slope of the gradient.

Figures 8.24 and 8.25 show plots of R_s versus $g^{-1/2}$ for the experimental results of Kato et al. (1982b, 1983a) and of Yamamoto et al. (1987c). The same dependence is seen as in Fig. 8.23. In the actual purification process of enzymes, R_s is not a measurable quantity owing to a number of overlapped peaks. Therefore, the purification factor PF defined by the following equation is often used as a measure of the degree of the purification:

$$PF = \frac{\left(\dfrac{\text{enzyme activity}}{\text{protein (or solid) concentration}}\right)\text{eluted fraction}}{\left(\dfrac{\text{enzyme activity}}{\text{protein (or solid) concentration}}\right)\text{sample}} \quad (8.9)$$

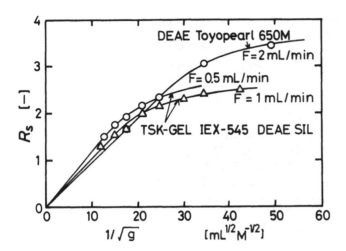

Fig. 8.24 Effect of the gradient slope on R_S for the experimental results of Kato et al. (1982b, 1983a). The details of the experimental conditions for a TSK gel IEX-545 DEAE SIL HPIEC column are the same as in Fig. 8.17. R_S of ovalbumin and trypsin inhibitor was measured on a DEAE-Toyopearl 650M MPIEC column (d_c = 1.6 cm, Z = 15 cm). (Experimental data are taken from Kato et al., 1982b, 1983a.)

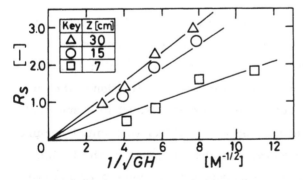

Fig. 8.25 Effect of the gradient slope on R_S. The experimental conditions are the same as in Fig. 8.20. (From Yamamoto et al., 1987c.)

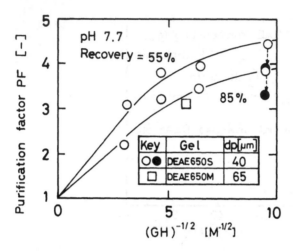

Fig. 8.26 Effect of GH on the purification factor PF as a function of recovery. The sample used is a crude β-galactosidase
from *Aspergillus oryzae*. The columns used were a DEAE-Toyopearl 650S MPIEC column (d_p = 40 μm, d_c = 1.6 cm, Z = 15 cm)
and a DEAE-Toyopearl 650M MPIEC column (d_p = 65 μm, d_c =
1.4 cm, Z = 15 cm). The linear gradient elution was performed
by a linear increase in NaCl concentration from 0.03 M in 14 mM
Tris-HCl buffer (pH 7.7) at 25°C. F = 2—3 mL/min. The sample (1% β-galactosidase) volume is (○ and □) 10 mL and (●) 50
mL. (Reproduced from Yamamoto et al., 1987b, by permission
of Elsevier Scientific Publishing Company.)

Figure 8.26 shows the relation between the PF of a crude en-

zyme and GH as a function of recovery (Yamamoto et al., 1987b).

The elution curve of this crude enzyme is shown in Figs. 7.2

and 8.37. The PF was calculated from the fractions, the sum

of the enzyme activities of which becomes equal to a specified

value (in this instance either 55 or 85%) of that initially applied

to the column. Therefore, considerable parts of the leading

and tailing portions of the enzyme activity versus time curve

were ignored for the calculation of PF when the recovery was

55%. There is a linear relation between PF and $(GH)^{-1/2}$ at

low $(GH)^{-1/2}$ values. Thus, the degree of purification increases
with decreasing slope of the gradient g. However, the slope of
the PF versus $(GH)^{-1/2}$ curve becomes shallow with a further
increase in $(GH)^{-1/2}$.

It is also seen from Fig. 8.26 that PF decreases with an increase in particle diameter, sample volume, and recovery at a
given GH and flow rate.

In general, it may be concluded that the resolution is improved by lowering the slope of the gradient. However, with
the decrease in g, which corresponds to the increase in $(GH)^{-1/2}$,
the separation time becomes long and the peak is diluted, as
shown in Sec. 8.1.3. For example, a twofold increase in
$(GH)^{-1/2}$ causes an almost fourfold increase in separation time.
Therefore, we must consider the relation among PF, separation
time, recovery, and amount of sample when choosing the operating conditions.

It should also be stressed here that, as described in Secs.
8.3.2 and 8.3.3, the effect of g and of Z on R_s is not independent but coupled. Therefore, we must choose these two parameters carefully so that the desired R_s can be obtained.

8.3.4 Particle Size

As mentioned in Sec 8.1.6, it is expected that the peak width
increases with the particle diameter d_p and the peak position is
independent of d_p if the ion-exchange capacity, the gel matrix,
and the other experimental conditions are the same. This was
clearly illustrated by Kato et al. (1983a). They compared chro-
matograms of a mixture of two proteins on DEAE-Toyopearl ion
exchangers of different particle diameters. Although the peak
positions of the two proteins are almost the same, the peak

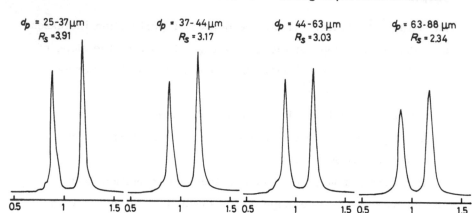

Fig. 8.27 Effect of particle diameter d_p on R_S. R_S was measured for a mixture of ovalbumin and trypsin inhibitor on DEAE-
Toyopearl 650 MPIEC columns ($d_c = 1.6$ cm, $Z = 15$ cm) or various particle sizes. (Reproduced from Kato et al., 1983a, by
permission of the authors and Elsevier Scientific Publishing
Company.)

width increases with d_p (Fig. 8.27). The purification factor is
also increased with the decrease in d_p, as shown in Fig. 8.26.

The simple relation between R_S and d_p is useful for scaling
up the separation. As described in Sec. 8.3.9, the R_S is proportional to d_p^{-1} at high flow rates. This is shown in Fig.
8.28. On the other hand, the R_S is proportional to $d_p^{-1/2}$ at
low flow rates. Kato et al. (1983a) have reported that R_S is
approximately proportional to $d_p^{-1/2}$.

As shown in Figs. 8.15 through 8.17, R_S^{-2} is a linear function of flow rate. Its slope is proportional to d_p^2, as seen in
Eq. (2.91b). Therefore, with decreasing particle diameter, the
flow rate dependence of R_S becomes weak (see Fig. 8.16). From
the preceding discussions, we can see that IEC columns packed
with small and rigid particles, such as high- or medium-perform-

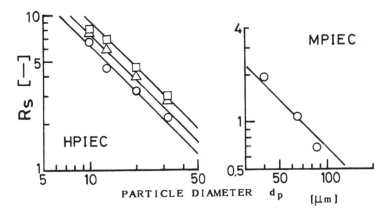

Fig. 8.28 Relation between R_s and d_p. The R_s values were measured for ovalbumin and trypsin inhibitor on a TSK gel DEAE 5PW HPIEC column (d_c = 0.75 cm, Z = 20 cm). The linear gradient elution was performed by a linear increase in NaCl in the buffer (20 mM Tris-HCl, pH 8.0). (△) F = 0.5 mL/min, g = 1.67 × 10^{-2} M/mL; (○) F = 1.0 mL/min, g = 8.33 × 10^{-3} M/mL; (□) F = 2 mL/min, g = 4.17 × 10^{-3} M/mL. The R_S values for β-lactoglobulins A and B on a DEAE-Toyopearl 650 MPIEC column (d_c = 1.6 cm, Z = 15 cm) were measured under the same experimental conditions as in Fig. 8.20 (u = 1.75 cm/min, GH = 0.032M). Note that the slope of the curves is 1.0.

ance IEC (MPIEC) give a high speed of separation and a high resolution (see also Fig. 7.2). For this reason, HPIEC can be used as a method for checking the homogeneity of the fractionated protein (Kato et al., 1985a; Yamamoto et al., 1987b). Figure 8.29 compares the elution curve on a HPIEC column for a crude enzyme with that for the peak fraction by MPIEC, shown in Fig. 7.2. The HPIEC method has the following advantages over conventional methods, such as disk gel electrophoresis: the speed of separation, the ease of the recovery, the reproducibility, and the quantitation (Kato et al., 1985a). These

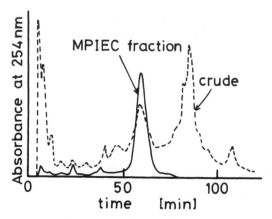

Fig. 8.29 Comparison of the elution curve of crude β-galactosidase on a HPIEC column with that of the purified fraction by
MPIEC. The linear gradient elution was performed by introducing a linear increase in NaCl concentration in the buffer to a
DEAE 5PW HPIEC column (d_p = 10 μm, d_c = 0.75 cm, Z = 7.5
cm) at 25°C. The initial buffer was 14 mM Tris-HCl containing
0.03 M NaCl; F = 0.5 mL/min, g = 0.0047 M/mL. The solid
curve is for the purified peak fraction by MPIEC shown in
Fig. 7.2. The dotted curve is for crude β-galactosidase. (From
Yamamoto et al., 1987b.) Reproduced by permission of Elsevier
Scientific Publishing.

characteristics also favor HPIEC as a method for monitoring

protein product formation during the fermentation process

(Gustafsson et al., 1986). HPIEC columns with nonporous pack-

ings that can resolve complex protein mixtures in 10 min have

been developed (Burke et al., 1986; Josic et al., 1986; Kato

et al., 1987). HPIEC is also useful for the rapid survey of the

optimum chromatographic conditions for large-scale MPIEC, when

the elution profiles between HPIEC and MPIEC are similar (Yam-

amoto et al., 1987b).

8.3.5 Temperature

The effect of temperature on the resolution of ovalbumin was in-

vestigated by Vanecek and Regnier (1980). They reported a

40% decrease in R_s as the temperature was lowered from 25 to 4°C. Similarly, Fig. 8.8 shows that R_s increases with temperature. In contrast, Frolik et al. (1982) found a decrease in R_s with increasing temperature that was caused by a change in the peak position. The data of Yamamoto et al. (1987c) are shown in Fig. 8.33. As discussed in Sec. 8.1.5, the temperature dependence of the adsorption equilibria is needed to predict the R_s value at a subambient temperature from that at near room temperature.

8.3.6 Mobile-Phase Composition (Salt, Buffer, and pH)

It is common to employ a simple salt, such as NaCl or KCl, in salt gradient elution. NaCl usually gives satisfactory results. However, the separation behavior was found to be affected by the type of salt. If the adsorption equilibrium between a protein and an ion exhcanger is governed only by simple equilibrium relations under the assumption that the activity coefficient is unity, (law of mass action; see Chap. 3), the ionic strength is a unique parameter that determines the adsorption equilibria. Therefore, as long as the ionic strength is the same, the elution by any salt will give a similar resolution. However, this expectation was found to be invalid (Kopaciewicz et al., 1983; Kopaciewicz and Regnier, 1983b; Regnier, 1984; Gooding and Schmuck, 1984). These studies have shown that the retention as well as the resolution of proteins was influenced by the type of salt. Kopaciewicz and Regnier (1983b) suggested that both a poor resolution and a low recovery can be improved by changing the type of salt. Unfortunately, neither systematic result nor qualitative explanation has yet been obtained for the relation between the type of the salt and the retention and resolution of proteins.

Preliminary experiments should therefore be carried out in order
to determine the salt that gives the highest resolution and recovery.

It should be borne in mind that the use of halide ions may
corrode the stainless steel surfaces employed in column tubes,
pumps, and other accessories. The general problem of corrosion
were discussed by Janson and Hedman (1982). Kato et al.
(1985a, 1985b) recommended the use of sodium perchlorate,
which is not very corrosive.

Various buffer systems were employed by Haff et al. (1983).
They reported that the R_S at a given pH often varied with the
buffer system by a factor of 2 and considered that this results
from a change in the width of the elution curve with the buffer
system. Volatile buffers may be useful since we can thus sim-
plify any subsequent process, such as dialysis (Frolik et al.,
1982).

IEC separation is based on the electrostatic interactions of
proteins and ion-exchange groups on IEC packings. Since the
net charge and the charge density of proteins vary with the
pH of the surrounding solutions, both the resolution and the
retention are also strongly dependent on the pH as well as on
the composition of buffer and salt solutions (Haff et al., 1983;
Gooding and Schmuck, 1985; Kopaciewicz et al., 1983; Richey,
1984). This is discussed again in Sec. 8.6.

8.3.7 Properties of the Stationary Phase

Vanecek and Regnier (1980) examined the effect of the pore
diameter of ion-exchange silica gels on R_S and peak position by
using gels with 10, 30, and 50 nm pore diameters. Both R_S

and the peak position were influenced by the pore diameter. The 30 nm pore diameter gel was found to have the highest loading capacity and R_s for proteins of molecular weight (MW) around 50,000. Gooding and Schmuck (1985) have reported that the resolution of a protein (MW = 140,000) is similar for both 30 and 100 nm pore packings, whereas smaller proteins are better resolved with 30 nm pore packings. As described in Chaps. 2 and 4, lowering the diffusion velocity of proteins in the pore of the packings causes an increase in the peak width. In order to prevent such lowering of the diffusivity, the pore diameters should thus be above 30 nm (Unger and Janzen, 1986). Most commercial ion-exchange packings for protein separation fulfill this requirement.

Kopaciewicz et al. (1985) have investigated the effect of the ligand density (ion-exchange capacity) and the hydrophobicity of IEC packings on retention and resolution. The resolution of ovalbumin and soybean trypsin inhibitor increased with increasing ligand density. However, the recovery of ferritin from the column decreased drastically with increasing ligand density. It also should be remembered that packings with a high ion-exchange capacity may change their volumes with the pH and/or salt concentration of the solution. The contribution of hydrophobic interaction in addition to electrostatic interaction to the retention of proteins onto the IEC column was also suggested by Kopaciewicz et al. (1985). The retention mechanism in IEC columns is discussed in Sec. 8.6.

8.3.8 Loading Capacity

Although the term "loading capacity in mg protein per mL column" carries no distinction between the sample volume and the

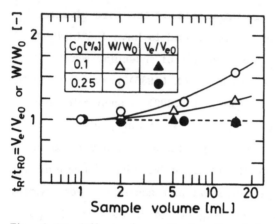

Fig. 8.30 Effect of sample (ovalbumin) volume on the peak position (t_R or $V_e = t_R F$) and on the peak width W. C_0 = ovalbumin concentration. W = width of the elution curve at the baseline (see Fig. 2.1). The $V_{e0} = t_{R0} F$ and W_0 values are those at sample volume = 1 mL. The linear gradient elution was performed by increasing the NaCl concentration in the buffer at 25°C. The initial buffer was 14 mM Tris-HCl (pH 7.7) containing 0.03 M NaCl. The column used was a DEAE-Toyopearl 650S (d_p = 40 μm, d_c = 1.6 cm, Z = 15 cm) MPIEC column. F = 2.8 mL/min, GH = 0.057 M, V_{e0} = 59.8 mL, W_0 = 10.2 mL. (From Yamamoto et al., 1987d.) Reproduced by permission of the American Institute of Chemical Engineers.

concentration, it is useful as a measure of sample loading   As shown in Sec. 8.1.2, when a sample protein obeys a linear isotherm, the sample volume can be increased until the sample is adsorbed to 20–30% of the column. Therefore, when the concentration of each protein is low and there is no interaction between the proteins, the sample volume can be applied according to these criteria without a significant loss of resolution.

As the protein concentration is increased, the peak position shifts to a smaller elution volume and the amount of tail increases, as shown in Sec. 8.1.4. This sample concentration overloading causes a decrease in R_s (for example, Kato et al.,

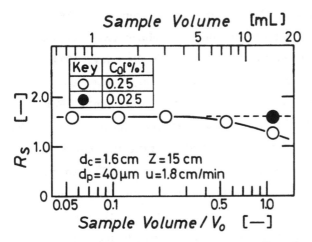

Fig. 8.31 Effect of sample volume on R_S of β-lactoglobulins A and B on a MPIEC column. The chromatographic conditions are the same as in Fig. 8.20. The column used is a DEAE-Toyopearl 650S (d_p = 40 μm, d_c = 1.6 cm, Z = 15 cm) MPIEC column. F = 1.5 mL/min, GH = 0.03 M. (From Yamamoto et al., 1987c.) Reproduced by permission of Elsevier Scientific Publishing.

1982b, 1983a; Nakamura and Kato, 1985; Vanecek and Regnier, 1980). However, the protein concentration below which R_S is constant is relatively high. This is because the linear part of the isotherm increases rapidly with ionic strengths as mentioned previously. Typical experimental results on the effect of sample loading on the peak position, the peak width, and R_S are shown in Figs. 8.30 and 8.31. These three quantities are constant up to a certain sample volume. A further increase in sample volume causes a considerable increase in peak width, a considerable decrease in R_S, and a slight decrease in peak elution volume. It is also noted that R_S is improved by lowering the initial sample concentration C_0 (Fig. 8.31). The maximum sample loading in Fig. 8.31 is 1.25 mg protein per mL column. Sample loading in analytical IEC separations is usually below

0.5 mg protein per mL column and that in preparative IEC separations is 0.5—4 mg protein per mL column (Kato et al., 1982a, 1983a, b; Nakamura and Kato, 1985; Regnier, 1984; Brewer et al., 1986; Scott et al., 1987). In some cases, the sample loading is above 10 mg protein per mL column. For example, the sample loading of a crude enzyme shown in Fig. 8.26 was increased from 3.3 to 17 mg crude enzyme per mL column with a small decrease in the purification factor. The sample loading in the linear gradient elution of two enzymes shown in Sec. 9.2.1 is 16 mg protein per mL column. In the plasma protein separation shown in Sec. 9.2.2, the sample loading is 30 mg protein per mL column. It should be noted that a sample loading of 1 mg protein per mL column is not as low as that in biospecific interaction affinity chromatography, in which only a protein of interest is adsorbed on the column and therefore the column is first saturated with the protein (Arnold and Blanch, 1986; Chase, 1984b; Clonis et al., 1986; Janson and Hedman, 1982).

The third overloading effect arises from the hydrodynamic instability of the viscous zone (Moore, 1970; Altgelt, 1970). Such an overloading effect is more pronounced when a short column of high efficiency is employed (Yamamoto et al., 1986a). Therefore, when the viscosity of the sample is different from that of the elution buffer by a factor of 2 or more, we should examine whether this overloading phenomenon occurs.

The fourth overloading effect is the displacement effect, as pointed out by Peterson (1970). However, if the displacement effect works like that in displacement chromatography, discussed in Sec. 8.5, it can separate substances. This is thus not a serious drawback.

In addition, if there is interaction between proteins, the situation becomes more complicated. Unfortunately, no simple method for predicting the effect of sample loading on R_s is available at present. Since sample overloading may be allowable in some cases in order to increase the throughput (Knox and Pyper, 1986; Dwyer, 1984), preliminary experiments should be carried out in order to examine the effect of sample loading on R_s.

To obtain a good resolution, the protein concentration should be reduced. In some cases a diluted sample gives a resolution equal to or better than that obtained with an original sample, as seen in Fig. 8.32 (see also Kato et al. 1982b). Moreover, dilution of the sample is useful for adjusting the pH and ionic strength to the initial starting condition (Scott et al., 1987). Therefore, we can disregard some processes for this purpose, such as dialysis and gel filtration chromatography. However, it should be kept in mind that proteins and enzymes are often unstable when they exist at low concentrations. In this case, a dilution of the sample is hazardous. Therefore, it is desirable to have information on the stability of proteins and enzymes to be separated before chromatography. Loading capacity of several types of liquid chromatography was discussed by Regnier and Mazsaroff (1987).

8.3.9 Productivity

In the design calculation of process IEC, we must consider the productivity. Let define the productivity P as

$$P = \frac{\text{amount of protein recovered}}{\text{mL column} \times \text{separation time}} \qquad (8.10)$$

at a certain purification factor PF or R_s. In order to calculate the productivity P, a simple parameter that relates R_s to the

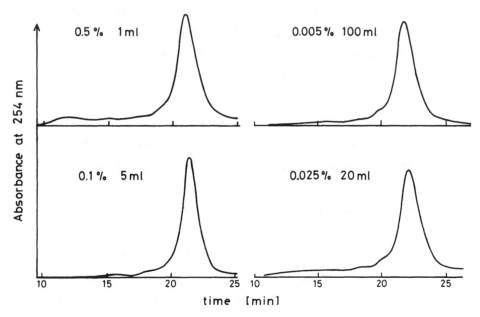

Fig. 8.32 Effect of sample load on the elution behavior. A DEAE-Toyopearl 650S MPIEC column (d_p = 40 μm, d_c = 1.6 cm, Z = 15 cm) was employed. The chromatographic conditions are the same as in Fig. 8.30. g = 3.4 × 10^{-3} M/mL, F = 3 mL/min. Note that the sample (five times crystallized ovalbumin) volume and concentration are chosen so that the equal weight of the sample is applied to the column.

combination of the column and operating variables is needed. When the dimensionless flow rate ν = ud_p/D_m is so high (for example, ν > 200) that the contribution of the A term is negligible to the total h = HETP/d_p in Eq. (2.47), HETP becomes

$$HETP \propto \frac{ud_p^2}{D_m} \qquad (8.11)$$

where D_m = molecular diffusivity. Insertion into Eq. (2.92) gives the relation (Yamamoto et al., 1987c)

$$R_s \propto Y = \left(\frac{D_m I_a Z}{GHud_p^2} \right)^{1/2} \tag{8.12}$$

where I_a = dimensional constant with a numerical value of 1. In Eq. (8.12), GH is employed instead of G since the I_R is a unique function of GH, as shown in Sec. 8.2.1. As is clear from Eq. (2.84) and the discussions in Sec. 8.2.1, θ_R (= $t_R u/Z$) is determined only by GH. The separation time $t_R = \theta_R Z/u$ is thus almost similar when $Z/(uGH)$ is the same. For example, when GH is decreased by a factor of 2, a twofold decrease in Z or a twofold increase in u gives the same t_R. (Although I_R increases with GH as shown in Sec. 8.2.1, for the sake of simplicity, it is assumed that the variation in I_R with GH is negligible.)

Figure 8.33 shows the relation between R_s and $Y = (D_m I_a Z/ GHud_p^2)^{1/2}$ for a wide range of experimental conditions (column length Z = 1−30 cm; particle diameter d_p = 40−87 μm; linear mobile-phase velocity u = 0.2−3.9 cm/min; column diameter d_c = 0.6−2.9 cm; temperature T = 15−35°C). It is seen that most experimental results are correlated with a single straight line. The results at low flow rates are lower than the straight line. This implies that the approximation of Eq. (8.11) is not valid at such low flow rates. The results with a column 1 cm in length become lower than the straight line in Fig. 8.33 with a decreasing slope of the gradient (the reason for this is described in Sec. 8.3.3). However, it is noteworthy that such short columns can separate proteins fairly well when proper chromatographic conditions are chosen, as seen in Fig. 8.33.

Figure 8.33 indicates that similar R_s values can be obtained at a given Y, although the value of each parameter included in the term Y is different. As mentioned earlier, the separation

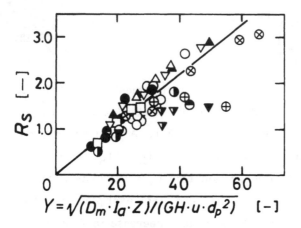

Key	d_p [µm]	d_c [cm]	Z [cm]	$\frac{u}{\text{[cm/min]}}$	T [°C]
⊕		0.6	1	1.6	
⊖				2.5	
◖				3.9	
◐	40		7	1.5	
○		1.6	15	1.8	
●				3.5	25
⊗			30	1.7	
▲		0.9	15	26	
△		1.4	15	1.8	
▽	65			1.8	
▼		2.9	15	0.5	
▼				0.2	
□	87	1.4	15	1.8	
⦶	40	1.6	15	1.7	35
⦵					15

Fig. 8.33 Relation between R_s and $Y = [D_m I_a Z/(GHud_p^2)]^{1/2}$. The chromatographic conditions are the same as in Fig. 8.20. (From Yamamoto et al., 1987c.) Reproduced by permission of Elsevier Scientific Publishing.

time is almost the same at a certain Z/(uGH). Thus the productivity of the experimental points located on the lower side
of the straight line in Fig. 8.33 is low. Therefore, the decrease in flow rate is not advantageous in increasing the productivity. The highest productivity is obtained when an IEC
column of short height packed with small particles is operated
at relatively high flow rates with a gradient slope that gives
a desired resolution. If we calculate the productivity P defined
by Eq. (8.10) for the linear gradient elution experiments of
crude β-galactosidase on high- and medium-performance IEC
columns shown in Fig. 7.2 and 8.26, the productivity, that is,
the P value of the HPIEC column, as summarized in Table 8.1,
is higher than that of the MPIEC column. The application to
large-scale MPIEC and HPIEC is shown in Sec. 9.2.

An increase in sample loading also increases the throughput,
although it lowers the resolution, as shown in Sec. 8.3.8. In
addition, the resolution also increases with a decrease in the
recovery, as shown in Fig. 8.26. Therefore, trial-and-error
calculation is needed in order to determine the chromatographic
conditions.

8.4 STEPWISE ELUTION

Stepwise elution, in which a discontinuous change in the elution
buffer is introduced into a column, is also frequently employed
instead of gradient elution in certain cases. This procedure
requires a much shorter operation time than gradient elution
and may be successfully applied to the separation of a mixture
of sufficiently different proteins. Some examples of applications
of stepwise elution were reviewed by Sober and Peterson (1958)

Table 8.1 Productivity of HPIEC and MPIEC Columns[a]

Column	d_p (μm)	d_c (cm)	Z (cm)	V_t (mL)	F (mL/min)	u_0 (cm/min)	g (M/mL)	t_s^b (min)	P^c (mg/ml min)
HPIEC	10	0.75	7.5	3.3	1.0	2.3	2.4×10^{-3}	45	0.0115
MPIEC	40	1.6	15.0	30.0	2.0	1.0	6.0×10^{-4}	108	0.00525

[a]The chromatographic conditions and the sample are the same as in Fig. 7.2. HPIEC column: TSK gel DEAE 5PW; MPIEC column: DEAE-Toyopearl 650S. The sample (1% crude β-galactosidase) volume was one-third V_t. The PF value according to Eq. (8.9) and the recovery in these two experiments were approximately 4.0 and 85%, respectively.

[b]The separation time t_s is defined as the time at the end of the enzyme activity versus the fractionated time curve.

[c]In the calculation of P by Eq. (8.10), it is assumed that 1 g crude enzyme contains 0.2 g pure enzyme. Note that the I_R value at a certain GH for the HPIEC column is higher than that for the MPIEC column. This means that the enzyme is more strongly retained on the HPIEC column. Therefore, if the HPIEC or MPIEC columns of different particle sizes are employed, a greater difference will be found in the P value.

and recently by Janson and Hedman (1982). Furthermore, the apparatus and operation procedure are simple compared with those for gradient elution. However, one should remember that there are several disadvantages to stepwise elution, as pointed out by several researchers (Morris and Morris, 1964; Saunders, 1975). One of the most serious disadvantages is an artificial peak due to a discontinuous change in an elution buffer. For example, if the ionic strength of the elution buffer is high enough for all the proteins contained in the sample to be desorbed completely (type I elution), all the proteins elute as a single peak. On the other hand, when the ionic strength of the first elution buffer is low (type II elution), the second elution buffer may cause an artificial peak, which is frequently called as a "false peak." These phenomena make it difficult to interpret the experimental results and may lead to a misunderstanding of the homogeneity of the eluted fraction. Of course, these artificial peaks can be avoided if the elution schedule, such as the salt concentration or pH of the elution buffer and its volume applied to the column is made properly. In order to make a good elution schedule, at least fundamental knowledge, shown in this section, is needed.

8.4.1 General Elution Behavior

Elution behavior in stepwise elution can be grouped into two types, as was qualitatively explained in Sec. 2.2.1. In type I, a protein is desorbed completely in an elution buffer, and therefore it is eluted in the spreading front boundary of the elution buffer. In type II, the protein peak appears after the concentration of the composition of the elution buffer at the exit of the column reaches its initial value.

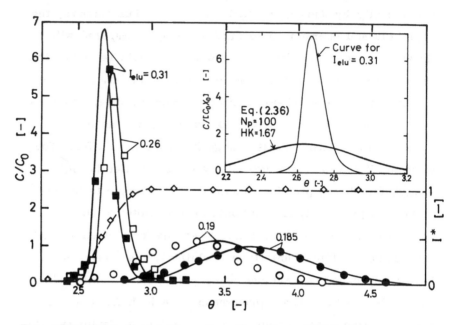

Fig. 8.34 Stepwise elution of ovalbumin on DEAE-Sepharose CL-6B with various ionic strengths of elution buffer. The solid curves are the calculated results for the protein with $N_p = 100$ for $I_{elu} = 0.31$ and 0.26 M, $N_p = 83$ for $I_{elu} = 0.19$ M, and $N_p = 80$ for $I_{elu} = 0.185$ M. Here, I* is the dimensionless ionic strength represented by $I^* = (I - I_0)/(I_{elu} - I_0)$. The calculated results with $N_p' = 400$ and experimental results ($\diamond$) for various I_{elu} are almost the same as when they are converted to I* by the preceding equation. Experimental conditions were $d_c = 1.5$ cm, $Z = 10$ cm; sample volume 6 mL; $C_0 = 0.02\%$; $I_0 = 0.11$ M, $F = 18$ mL/h; pH 7.9 and temperature 20°C. (From Yamamoto et al., 1983b.)

These two types of elution behavior can be clearly under-stood from Fig. 8.34, where experimental results for the step-wise elution of ovalbumin on DEAE-Sepharose CL-6B (Yamamoto et al., 1983b) are shown. The stepwise elution was performed by an increase in NaCl concentration in the elution buffer. As shown in the figure, the shape and peak position of the elution curves are markedly influenced by a small change in the ionic strength of the elution buffer I_{elu}. The elution curves with I_{elu} = 0.185 and 0.19 belong to type II. The elution curve becomes wider and the peak position shifts to a larger θ value with the decrease in I_{elu}. The peak position may be predicted by using the equation

$$\theta_R = 1 + HK_{elu} \tag{8.8}$$

where θ_R = θ value at the peak position and K_{elu} = K value at $I = I_{elu}$. Therefore, theoretical curves were calculated with the N_p values determined using the relation $K_R = K_{elu}$. The peak width is also predictable by Eq. (2.75) in Sec. 2.2.2 by using the relation $N_p = (\theta_R / \sigma_\theta)^2$.

Type I elution behavior is found in the elution curves for I_{elu} = 0.26 and 0.31. Both elution curves are similar and much sharper than those for type II and have a sharp front and a short tail. It also should be noted that the protein concentration at the peak position is much higher than that initially applied to the column. In other words, the protein can be recovered as a concentrated fraction. The K_{elu} value for this case is smaller than K'. The protein zone then moves with the spreading boundary of the elution buffer during the elution process. In this case R_f = moving velocity of zone/moving velocity of the elution buffer = 1. This elution mechanism in type I is quite similar to the elution of a steep gradient. As described in Sec. 8.2.1, K_R for the steep gradient elution becomes K'. The K_R values for these elution curves calculated from the measured I_R were almost equal to K'. The theoretical elution curves are calculated with N_p = 100 determined by K_R = K' and N'_p = 400. In this case, the effect of N'_p is significant since the shape of the spreading boundary of the elution buffer plays a role similar to that of the gradient slope in linear gradient elution. In order to examine the zone sharpening effect, an elution curve was calculated by Eq. (2.36), in which the zone sharpening effect is not taken into consideration, and compared with the curve for $I_{elu} = 0.31$. In this calculation, the same N_p and K_R values were used as those employed for calculating the elution curve for $I_{elu} = 0.31$ by the rigorous model. It is clear from the inset of Fig. 8.34 that the elution curve by the rigorous model is much sharper than that calculated by Eq. (2.36). This drastic reduction in peak width is due to the zone sharpening effect.

The effect of the sample volume in type I elution is shown in Fig. 8.35. It is seen that the peak position and shape is not dependent on sample volume. At a sample volume of 12 mL, the maximum concentration is nine times as high as the initial concentration.

On the other hand, the effects of operating and column variables in type II elution are similar to those in isocratic elution, since the protein desorbed by the elution buffer is always subjected to it during the elution (i.e., $R_f < 1$). Therefore, the peak position and peak shape can be predicted in terms of the theories presented in Sec. 2.1. The only difference is that in isocratic elution the protein eluted never exceeds its initial concentration but in type II stepwise elution the protein may be concentrated.

8.4.2 False (Artificial) Peak

In the previous section, the elution behavior in stepwise elution was described, and Fig. 8.34 illustrated the effect of some of the operating variables. In the actual separation processes, step-

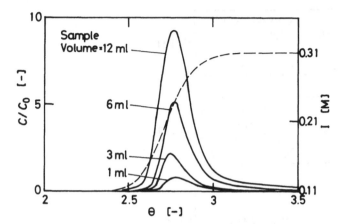

Fig. 8.35 Effect of sample volume in stepwise elution of type I behavior. Column, DEAE-Sepharose CL-6B (d_c = 1.5 cm, Z = 10 cm). Sample, bovine serum albumn. Flow rate F = 20 mL/h. The other conditions are the same as in Fig. 8.34. (From Yamamoto et al., unpublished data.)

wise elution is performed not by a single step but by a multistep elution procedure to fractionate several components contained in the sample. If each of the components in the sample is eluted as type I elution, the peak position can be readily predicted since the sample is eluted as a sharp peak in the spreading front boundary of each elution buffer. This condition can be accomplished only when the components of the sample have very different properties.

On the other hand, in type II elution, the elution mechanism becomes rather complicated. The peak that moves with a velocity defined by the first elution buffer changes its velocity after it has traveled some distance since the second elution buffer catches up with the protein peak. The peak thus moves with a velocity defined by the second elution buffer. This situation is illustrated in Fig. 8.36.

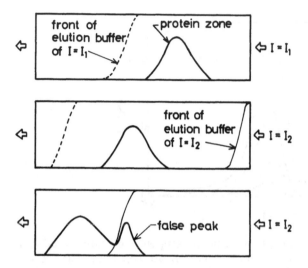

Fig. 8.36 Schematic representation of the occurrence of the artificial (false) peak.

An artificial peak, often called a false peak, appears in stepwise elution. When the protein moves slowly with an elution buffer, that is, a large distribution coefficient, the protein zone spreads as predicted by Eq. (2.75). If the next elution buffer by which the protein is considerably desorbed reaches the rear boundary of the protein zone, it causes an artificial peak, as shown in Fig. 8.36. This is the so-called false peak, which is confusing in testing the homogeneity of the sample.

8.4.3 Application to Enzyme Purification

As explained, it is necessary to know the dependence of the distribution coefficient of proteins on the salt concentration of the pH in order to obtain good resolution by stepwise elution. Since no specific interaction is expected between a desired protein or enzyme and an IEC column, we must seek the salt concentration

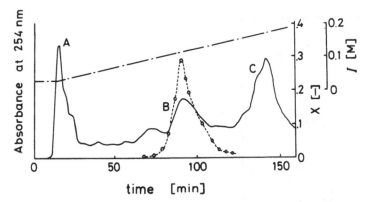

Fig. 8.37   Elution curve of crude β-galactosidase on a MPIEC column.   The linear gradient elution experiment was performed by a linear increase in NaCl concentration in the buffer (14 mM Tris-HCl, pH 7.7) from 0.03 M at 25°C.   The column used was a DEAE-Toyopearl 650S MPIEC column (d_p = 40 μm, d_c = 1.6 cm, Z = 15 cm).   The slope of the gradient g = 6 × 10^{-4} M/mL, F = 2 mL/min.   Sample (crude β-galactosidase), 1%, 10 mL.   X is the ratio of the enzyme activity to that of the sample.   (- ·-), I; (--○--), X.   (From Yamamoto et al., 1987e.)

I and the volume of the elution buffer such that the desired protein is retained and the contaminants are eluted (or vice versa).   Therefore, it is necessary to know the salt concentration dependence of the distribution coefficient K(I) of the contaminant proteins as well as of the desired protein.   However, measurement of K(I) is time consuming and laborious, since a number of contaminant proteins are included in the actual starting crude sample, for example fermentation broths.

A method for determining the chromatographic conditions in stepwise elution is briefly described here.   The elution curves of the sample used (crude β-galactosidase) in the linear gradient elution on medium- and high-performance ion exchange columns were shown in Fig. 7.2.   The curve on the MPIEC column is again shown in Fig. 8.37.   Although there are a number of peaks,

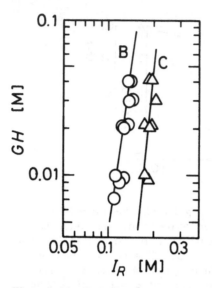

Fig. 8.38 Relation between GH and the ionic strength at the
peak position I_R for peaks B and C in Fig. 8.37. The linear
gradient elution experiments were carried out with various slopes
g of the gradient under the experimental conditions shown in
Fig. 8.37. The ionic strength = NaCl molarity + 0.01 M (= the
buffer ionic strength) at the peak position I_R was then measured.
GH = $g(V_t - V_0)$, where V_t = total column volume and V_0 =
column void volume. Peak B is the enzyme activity; peak C
is the contaminant (see Fig. 8.37). (From Yamamoto et al.,
1987e.)

including a peak with enzyme activity, we focus our attention

on the three large peaks A, B, and C shown in Fig. 8.37.

Peak A is eluted at very low salt concentrations. Therefore,

the chromatographic conditions should be such that peak B, the

desired enzyme, is eluted whereas peak C is retained. The GH

versus I_R plots for peaks B and C are shown in Fig. 8.38.

The K versus I relations obtained from differentiation of these

curves on the basis of Eq. (2.85) are shown in Fig. 8.39 (see

also Sec. 8.2.1). From these results, the ionic strength of the

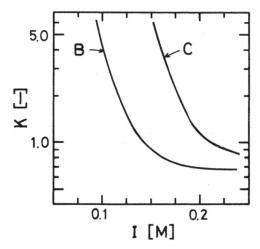

Fig. 8.39 Relation between the distribution coefficient K and ionic strength I. The solid curves were obtained from the curves in Fig. 8.38 on the basis of the method shown in Sec. 8.2.1. (From Yamamoto et al., 1987e.)

elution buffer I_{elu} was determined as 0.16. After washing the column with the initial buffer, this elution buffer was applied. The elution curves obtained with different sample volumes are shown in Fig. 8.40. In order to verify the homogeneity of the fractions in Fig. 8.40, linear gradient elution on a HPIEC column was carried out with the fractions as a sample. As shown in Fig. 8.41, although peak A is removed, peak C was included in both fractions. The ratio of the area of peak C in the fraction increases with sample volume. This corresponds to the decrease in the purification factor PF defined by Eq. (8.9). The PF values are lower than those in linear gradient elution shown in Fig. 8.26 and are presented in Sec. 10.2.

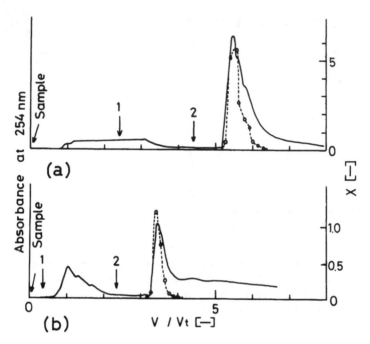

Fig. 8.40 Elution curves of β-galactosidase in stepwise elution.
V = volume from the start of the experiment, V_t = total column
volume, X = ratio of enzyme activity to that of the sample. The
sample (1% crude β-galactosidase) dissolved in the starting buffer (14 mM Tris-HCl, pH 7.7, containing 0.03 M NaCl) was
applied to a DEAE-Toyopearl 650S MPIEC column (d_p = 40 μm,
d_c = 1.6 cm, Z = 15 cm). The sample volume was (a) 75 mL
and (b) 10 mL. The arrows in the figure indicate the change
in elution buffer: (1) the starting buffer (washing); (2) 14
mM Tris-HCl, pH 7.7, containing 0.15 M NaCl (elution). Flow
rate F = 1.5 mL/min, 25°C. The purification factor defined by
Eq. (8.9) is (a) 1.7 and (b) 2.2. The recovery was 92–93%
for both cases. (--○--), X. (From Yamamoto et al., 1987e.)

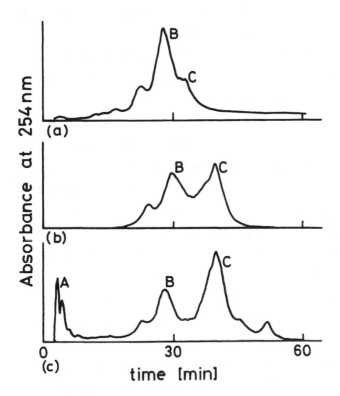

Fig. 8.41 Comparison of the elution curves of the purified fraction by the stepwise elution shown in Fig. 8.40 with crude β-
galactosidase on a HPIEC column. The chromatographic conditions for the linear gradient elution with a TSK gel DEAE 5PW
HPIEC column are the same in Fig. 8.29. The sample used is
(a) the fraction in Fig. 8.40b; (b) the fraction in Fig. 8.40a;
(c) crude sample. F = 1 mL/min, g = 0.0047 M/mL.

8.5 OTHER SPECIAL SEPARATION METHODS
AND APPARATUS

In addition to gradient and stepwise elution by a change in the

salt concentration (ionic strength), several elution techniques

have been employed in the ion-exchange chromatography of pro-

teins. Although it is not common to employ an isocratic elution

method in the IEC of proteins, some examples of the applications
of this method are found in the literature. The isocratic elution method is successful when mass transport between the mobile
and stationary phases is rapid. This condition can be satisfied
only when proteins are adsorbed on the surface of the ion exchanger and are not accessible to the interior of the ion exchanger. Therefore, application of isocratic elution is restricted
to the resin ion exchanger, highly cross-linked ion-exchange
gels, or cellulosic ion exchangers. Many of the early attempts
to separate proteins by IEC were performed using isocratic elution on resin ion exchangers. However, as pointed out by
Fasold (1975), special attention must be paid to the denaturation
of proteins owing to the strong binding between proteins and
the resin ion exchangers. Kirkegaard (1972; 1976) proposed an
isocratic elution method with highly cross-linked dextran ion exchangers and used the term "ion filtration chromatography." In
this method, a protein applied to the column has a weak electrostatic interaction with ion-exchange groups at the surface of
the gels and is eluted within approximately one column volume.
In other words, the distribution coefficient of the protein is less
than 1.0. Figure 8.42 is a schematic illustration of the conditions for ion filtration chromatography. Since a small change in
ionic strength or pH of the elution buffer influences the distribution coefficient, as shown in Fig. 8.42, the composition of
the elution buffer must be chosen carefully. If the distribution
coefficient is available as a function of the ionic strength and
pH, the choice of the elution conditions is easier. Kirkegaard
(1973; 1976) and others (Johnson, 1974; Johnson and Bock,
1974) also used highly cross-linked dextran ion exchangers for
the gradient elution of proteins. It should be mentioned that

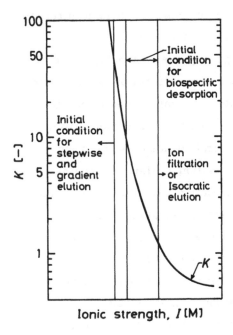

Fig. 8.42 Schematic relation between K and I and initial conditions suitable for several elution methods.

some proteins are able to diffuse even inside such highly cross-linked ion-exchange gels. This results in a highly asymmetric elution curve with a long tail, as mentioned previously. Ruckenstein and Lesins (1986) have presented an isocratic elution method for proteins that uses ion-exchange columns. This method, which they refer to as "potential barrier chromatography" (PBC), is based on the high sensitivity of the interaction potential between proteins and the stationary phase to small differences in their electrical charges and hydrophobicity. The sign of the charge of the stationary phase and of proteins are the same in PBC, although they are opposite in the usual IEC. In order to increase the retention of proteins, a change in the mobile-phase composition, such as an increase in ionic strength, is needed.

Recently, "affinity chromatography," a chromatographic separation method that utilizes biospecific adsorption between proteins and ligands introduced into stationary phases, has been rapidly developed. Although we cannot expect such strong biospecific adsorption in IEC, a protein adsorbed onto ion exchangers may be eluted with an elution buffer that contains a component with a biospecific interaction with the protein and the same charge as that of the ion exchanger. Scopes (1977a, 1977b) purified a number of enzymes by using the biospecific desorption method. According to his reports, the advantage of this method is that it is not necessary to prepare the gel with biospecific ligands and ordinary ion-exchange gels can be employed. The disadvantages are as follows. For elution, the polarity of the charge of the ion exchangers must be the same as that of the ligands and opposite to that of the enzymes. If the ratio of a desired enzyme to the total protein solution is small, this method is not suitable. In addition, enzymes must be adsorbed weakly so that the elution can be performed with a low ligand concentration. This implies that the initial conditions suitable for this method are very limited. Yon (1980) investigated factors affecting biospecific desorption by using a computer simulation. His computed results agree with the results of Scopes. In Yon's model, three adsorption equilibria are considered: protein and ligand, protein and adsorption site, and protein-ligand complex and adsorption site. He concluded that the equilibrium association constant between protein and adsorption site must be in the range 10^6-10^7 M^{-1}. A possible range for the initial condition of the biospecific desorption

method is also shown schematically in Fig. 8.42. The isocratic
elution method, as well as the biospecific desorption method,
utilizes a very small part of the ion-exchange capacity although
current ion exchangers have a high protein adsorption capacity.

On the other hand, in another chromatographic method, "displacement chromatography," the protein concentration in the ion
exchanger is very high, sometimes maximum for a given substance. The utility of displacement chromatography was first
claimed by Tiselius (1943). The principle of displacement chromatography is illustrated as follows with the aid of Fig. 8.43.
First let us consider the displacement of substance 3 by substance 4 in Fig. 8.43. Substance 4 has the strongest affinity to
the ion exchanger and moves in the column with velocity = $u/(1
+ HK)$, where $K = \overline{C}/C$ is the distribution coefficient at $C = C_4$
when the initial concentration of substance 4 is C_4. (For the
meaning of other variables, see Sec. 2.1.2.) As discussed in
Sec. 2.1.2, the zone of substance 3 has a long tail due to non-

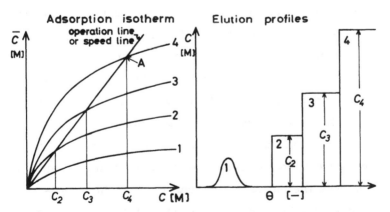

Fig. 8.43 Schematic drawing of elution profiles in displacement chromatography and graphic determination of the concentration of each component from adsorption isotherms.

linearity of the isotherm, which is convex upward (favorable), as
shown in Fig. 8.43. However, since substance 3 cannot be adsorbed to the ion exchanger in the presence of substance 4,
such a long tail is expelled by substance 4 and consequently the
zone is compressed. In other words, this causes an increase
in the moving velocity of the rear boundary of the zone of substance 3 compared with that of the front boundary. After a
certain time, a steady state is attained in which the moving
velocity of substances 3 and 4 is the same. The concentration
C_3 of substance 3 is adjusted automatically so that $K = \overline{C}/C$ for
substance 3 at $C = C_3$ is equal to K of substance 4 at $C = C_4$.
The width of zone for substance 3 is also determined from a
mass balance such that a mass of substance 3 contained in the
zone must be the same as that originally in the sample solution.
Similarly, the concentration and width of zone for substance 2
with a weaker affinity than that of substance 3 is determined
automatically. A graphic determination of the concentration of
each substance during steady state is shown in Fig. 8.43,
which Tiselius first proposed. As shown in the figure, the concentration of each substance can be determined from the intersection of the straight line drawn from the origin to point A
(called a speed line or an operation line) and the respective isotherm. It should be noted that there is no intersection between
the operation line and the isotherm of substance 1. In this
case, elution of substance 1 is similar to isocratic elution and
the elution curve appears before the front boundary of substance 2 reaches the outlet of the column. In other words, it
is a necessary condition for displacement chromatography that the
isotherm of each substance have an intersection with the operation line. As shown in Fig. 8.43, a series of elution profiles

with a sharp front and boundary can be obtained at the outlet of
the column. Clearly, when displacement chromatography works
properly, it is an efficient method since substances can be separated in a concentrated form without contamination. However,
displacement chromatography was not until recently employed for
the separation of proteins since there was no suitable displacer.
Leaback and Robinson (1975) used Ampholine ampholytes as
displacers for the separation of two isoenzymes. Several applications of this method are found in the literature (Young and
Webb, 1978; Young et al., 1978). Peterson and Torres and
their coworkers (Peterson, 1978; Peterson and Torres, 1983,
1984; Torres et al., 1984, 1987) have reported the displacement
chromatography of proteins using specially prepared carboxy-
methyl dextrans (CM-D) as displacers. It is claimed that the
displacement chromatographic technique is useful because the
fractionated sample has a very low salt concentration and is directly applicable to gel electrophoresis. It is also stressed that
this method has a much higher sample loading capacity (for example, 85 mg protein per mL column with specially fractionated
CM-D and 8 mg protein per mL column with unfractionated CM-D)
than that of the usual elution methods (see Sec. 8.3.8).

Most applications of protein IEC are by nonisocratic elution
by increasing ionic strengths at a fixed pH. This is because
the operation is simple: the ionic strengths can be increased
simply by increasing the salt concentration in the elution buffer.
However, the elution is sometimes performed by a change in pH.
One of the problems with pH gradient elution is that it is difficult to make a continuous pH-gradient using a simple apparatus
owing to the buffer action of the elution buffer and of the ion
exchanger. A small stepwise increase in pH was employed by

Lampson and Tytell (1965) for estimating the isoelectric points of proteins. Boman (1955) produced a continuous pH gradient by using an open mixing chamber and a vessel with a conical section, and mentioned that a buffer system with about the same buffering capacity over the region of the gradient must be employed to produce a good pH gradient. Huisman and Dozy (1965) produced a continuous decreasing pH gradient by a variable gradient device composed of nine vessels, each of which was filled with a buffer of different pH values. They also showed a good correlation between pH at the peak position and the relative mobility in starch-gel electrophoresis. Bardsley and Wardell (1982) presented a method for producing non-linear pH gradients. Chromatofocusing, introduced by Sluyterman and Elgersma (1978) and Sluyterman and Wijdenes (1978), may also be considered elution by pH gradient, although the pH gradient is produced internally in the column during elution. In pH gradient elution, one may expect that a protein is eluted at the pH near its isoelectric point. This is one advantage to this elution technique. However, since the elution mechanism in pH gradient elution is similar to that in salt gradient elution, care must be taken for the shift in pH at which the peak is eluted with a change in the slope of the gradient caused by the mechanism described in Sec. 8.2.1. Kopaciewicz et al. (1983) examined the retention behavior of proteins on ion exchangers and found that some enzymes were retained on the ion exchangers even at their respective isoelectric points. Furthermore, if the pH gradient elution or chromatofocusing technique is employed for the determination of the isoelectric point of proteins, the shift in pH in ion exchangers from that in the solution due to the Donnan potential must be taken into consideration, as pointed out by Sluyterman and Elgersma (1978). Adachi et al. (1978)

proposed a method for differentiating the pH in the ion exchanger
from that in the outer solution by assuming Donnan's equilibrium
(see also Chap. 3).

In general, the elution behavior in pH gradient elution may
be predicted in the same manner as that in salt gradient elution
when the distribution coefficient as function of pH and protein
concentration and the HETP at the pH where the protein is not
adsorbed to ion exchangers are known.

As stated in the introductory chapter, chromatography is
essentially a batch separation method. Thus, for the automation
of ion-exchange chromatographic separation processes, a cyclic
operation that consists of equilibration, adsorption, desorption
(elution), and washing in series, must be designed (McCoy,
1985a). Recent developments in microprocessors, analog-digital
or digital-analog converters, and several electrical interfaces
make it possible to easily construct such an automated cyclic
operation apparatus (Chase, 1984a; 1985). In fact, several
preparative liquid chromatography systems in which a microcomputer is installed are now commercially available.

Two types of chromatography have already been reported as
continuous or semicontinuous chromatography. The first is an
annulus column packed with gel filtration chromatography media
(Fox et al., 1969). In this system, while the column rotates at
a constant rotational speed, a sample is continuously fed at a
fixed point at the column inlet. Substances contained in the
sample travel down the rotating column with different trajectories,
which are governed by the respective distribution ceofficients
of the substances. The separated substances can be continuously collected at fixed points at the outlet of the column. The
separation and yields are as good as those with the usual column.
However, it seems that the application of this apparatus to non-

isocratic elution is difficult. This method was further analyzed
by several researchers (Begovich and Sisson, 1984; Arnold et
al., 1985b, and references cited therein).

On the other hand, a simulated moving bed (SMB) system
(Barker and Chuah, 1981; Barker and Thawait, 1986) may be
employed for a separation procedure that includes both adsorption and desorption steps. In this system, a moving port is
utilized so that continuous countercurrent chromatography can be
achieved. Hashimoto et al. (1983) proposed mathematical models
for designing this separation system. The SMB system can be
employed for rough separation by stepwise elution. Application
of the SMB system to the affinity chromatographic separation of
proteins was reported (Huang et al., 1986). However, nonchromatographic zone spreading in the rotational port must be
eliminated to obtain a fine separation of multicomponents com-
parable to that obtained with a usual column. Wankat (1977b)
proposed a moving port chromatography in which the feed injection point is moved to improve the throughput. This method
is further studied by several researchers (Wankat, 1984; McCoy,
1985b). Centrifugal chromatography (Scott et al., 1974, and
references cited therein) may also be one of the methods for
increasing sample throughput. Scott et al. (1974) built a centrifugal gel filtration chromatography system with eluate monitoring. In centrifugal chromatographic separation, the liquid flow
is directed in the radial direction from the center of the disk.
On the other hand, a method has recently been developed in
which the liquid flow supplied by a pump is introduced from
the whole outer surface of the annulus to the center (Peacock,
1986). Coiled sheets containing ion-exchange groups are employed as an annulus bed. This method is also aimed at the

large-scale purification of proteins.

The concept of two-dimensional separations is that a sample is subjected to two separation modes oriented at right angles to one another (Giddings, 1984; McCoy, 1986; Wankat, 1977a). Consequently, the two-dimensional method gives a high resolution. However, a two-dimensional method including IEC has not yet been reported.

Parametric pumping is also an efficient method for preparative separation (Wankat, 1974b). Its application to protein separation with IEC columns was reported previously (Chen et al., 1977). Hollein et al. (1982) used the electrical field in addition to the pH as a variable in the parametric pumping separation of proteins on ion-exchange columns.

A magnetic field was considered to stabilize the packed or fluidized bed (Burns and Graves, 1985; Siegel et al., 1986).

A column switching technique (Kopaciewicz and Regnier, 1983a) is also effective in increasing the throughput. Yamamoto et al. (1986b; 1987d) separated a mixture of three proteins (β-lactoglobulins A and B and ovalbumin) with close isoelectric points using the column switching technique. Figure 8.44 is a block diagram of the apparatus and the separation sequence used in this study. As shown in Fig. 8.48, both β-lactoglobulins A and B are adsorbed to a cation-exchange chromatography (CEC) column at pH 5.2 and eluted as a single peak, whereas ovalbumin is not adsorbed to the CEC column. On the other hand, β-lactoglobulins A and B are separated effectively on an anion-exchange chromatography (AEC) column at pH 5.2. On the basis of these results, the following procedure was debices. In the first step, ovalbumin, which is not adsorbed to the CEC column, was recovered. The second step is the desorption of β-lacto-

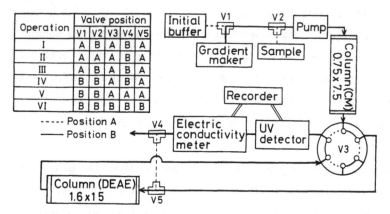

Operation	Valve position				
	V1	V2	V3	V4	V5
I	A	B	A	B	A
II	A	A	A	B	A
III	A	B	A	B	A
IV	B	B	A	B	A
V	B	B	A	A	A
VI	B	B	B	B	B

----- Position A
——— Position B

Fig. 8.44 Block diagram of the column switching technique. (From Yamamoto et al., 1987d.) Reproduced by permission of the American Institute of Chemical Engineers.

globulins A and B from the cation column. Since the CEC column is small and the slope of the gradient is shallow, the ionic strength at which the peak is eluted is low. Consequently, this fraction is directly introduced to the AEC column without the desalting process. In this third step, β-lactoglobulins A and B are separated on the AEC column. It should be noted that the sample is always subjected to the elution process until it is collected as a purified fraction. The experimental elution curves are shown in Fig. 8.45. Resolution of the three proteins separated by this process was far better than that with the single-step process shown in Fig. 8.46. In addition, the separation time was 140 min, which was 70% of the single-step separation time. These results demonstrate the versatility of the column switching technique. However, in order to obtain successful results by this method, it is necessary to know the peak position (retention time). For this reason, the method presented in Sec. 8.2.1 is recommended.

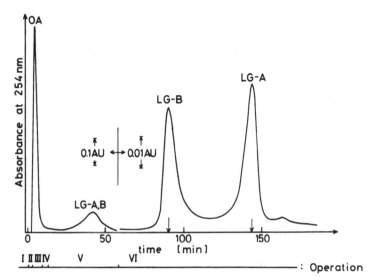

Fig. 8.45 Elution curves in the separation experiment using the column switching technique. Sample: 0.25% ovalbumin (OA) + 0.25% β-lactoglobulins A (LG-A) and B (LG-B), 2.5 mL. Roman numerals I—VI correspond to those in Fig. 8.44. The arrows represent the peak positions estimated from Fig. 8.48. The linear increase in NaCl concentration ($g = 4 \times 10^{-4}$ M/mL) was applied at a flow rate of 1.2 mL/min. The initial buffer was 10 mM acetate buffer (pH 5.2) containing 0.03 M NaCl. The GH [$= g(V_t - V_0)$] value was 0.00069 M for the CM Toyopearl 650S column (d_p = 36 μm, d_c = 0.75 cm, Z = 7.5 cm) and 0.0073 M for the DEAE-Toyopearl 650S column (d_p = 40 μm, d_c = 1.6 cm, Z = 15 cm), respectively. AU = absorbance unit. (From Yamamoto et al., 1987d.) Reproduced by permission of the American Institute of Chemical Engineers.

In the preceding application, the CEC and AEC columns are connected in series, although the ovalbumin fraction is drawn from the outlet of the CEC column. When the combination of AEC and CEC columns ("tandem columns") or a column packed with both AEC and CEC packings ("mixed-bed columns") is used, the separation of certain proteins may be improved (Josic et al., 1986; El Rassi and Horvath, 1986).

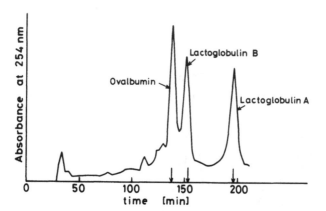

Fig. 8.46 Elution curves in linear gradient elution with a
DEAE-Toyopearl 650S MPIEC column. Sample: 0.5% ovalbumin +
0.5% β-lactoglobulins A and B, 2 mL. Column: DEAE-Toyopearl 650S (d_p = 40 μm, d_c = 1.6 cm, Z = 30 cm). The initial
buffer: 10 mM acetate buffer (pH 5.6) containing 0.03 M NaCl.
g = 3.8×10^{-4} M/mL, F = 1.5 mL/min. The arrows represent
the peak positions estimated from Fig. 8.48. (From Yamamoto
et al., 1987d.) Reproduced by permission of the American
Institute of Chemical Engineers.

Although IEC is usually employed in the later stage of the
purification process (Dunnill and Lilly, 1972; Bonnerjea et al.,
1986), it can be used in the early stage as a method for the
concentration or recovery of proteins or for the removal of contaminants. For this purpose, a short column is first saturated
with a sample and then elution is performed with a high-ionic-
strength buffer. This operation can also be carried out without
a packed bed. For example, ion exchangers are introduced in a
solution containing proteins, such as fermentation filtrates. The
ion exchangers that adsorbed the proteins are gathered in a filter and washed with a high-ionic-strength buffer to desorb the
adsorbed proteins. This type of method is applied to the large-
scale isolation of factor IX from plasma (Stampe et al., 1986).

In this application, 1.5 g of ion exchange gel (DEAE-Sephadex A-50) per kg plasma was used for 120 kg of plasma. Bruton et al. (1975) used 14 kg of dry cellulose ion exchangers (DEAE-cellulose DE-23) for the recovery of proteins from cell extract (95 L). The ammonium sulfate precipitation conventionally employed for protein recovery at the beginning of the protein purification process (Charm and Matteo, 1971; Scopes, 1982) was omitted in these applications. When the amount of the undesirable contaminant protein is much lower than that of the desirable protein, it is advantageous to adsorb the undesirable protein rather than the desirable protein to the ion exchanger. This type of operation is sometimes called "negative adsorption." Yang et al. (1987) purified heparinase by the negative adsorption method with a strong anion exchange column (QAE-Sephadex A25). Ayers (1985) presented a method for removing glucosyl-transferase, which catalyzes the synthesis of nonfermentable carbohydrates, from crude glucoamylase. The crude culture filtrate was directly introduced to the cation-exchange column, which adsorbs glucosyltransferase and does not adsorb gluco-amylase, until the column was saturated with glucosyltransferase. Although this method was carried out with the column, Ayers (1985) suggested the use of a stirred batch method, which is much simpler. When the ion exchanger is soft and cannot withstand agitation, a fluidized bed may be employed (Porath, 1972).

Owing to their high protein adsorption capacity, ion exchangers can be employed as a method for the concentration of proteins. Miller et al. (1976) employed a gel filtration column (d_c = 2 cm and Z = 90 cm) on top of which a small amount of ion-exchange gels is layered (d_c = 2 cm and Z = 3 cm) for the separation of proteins present in a very diluted solution. Nearly

1 L of enzyme solution was applied to this column, which was first adsorbed to the ion-exchange layer. The elution with the high ionic strength buffer was then started. The desorbed concentrated sample was sequentially introduced into the gel filtration column. The resulting resolution was as good as that obtained with the sample concentrated by pressure dialysis before chromatography. Miller et al. (1976) also stated that concentration using ion exchangers is useful since several enzymes, such as membrane-bound enzymes, require the presence of detergents, which makes it difficult to employ the usual concentration methods, such as ultrafiltration and salt precipitation. Rhodes et al. (1958) reported that 500 mL of conalbumin solution (0.1%) was concentrated by a factor of nearly 20 by a batch adsorption-desorption method with carboxylmethyl ion-exchange celluloses. In Table 8.2, the concentration of a protein (ovalbumin) on a DEAE IEC column is summarized. The ovalbumin solution (0.02%) was applied to the column equilibrated at pH 7.7 and ionic strength = 0.04 until the column was saturated with the solution. Subsequent desorption with the buffer containing 0.3 M NaCl yielded a 76-fold concentrated fraction.

In order to separate hydrophobic or membrane proteins by IEC, mobile-phase additives that can increase the solubility of the proteins are needed (Regnier, 1984). Since ionic solubilizing agents at high concentrations may interfere with the retention of proteins on the IEC column, the use of nonionic surfactants or organic solvents is recommended (Tandy et al., 1983; Ikigai et al., 1985).

Usually, the sample is adjusted to the initial condition, such as pH and salt concentration, by dialysis or gel filtration before it is applied to the IEC column. Vardanis (1985) found that the

Table 8.2 Concentration of Ovalbumin Solutions[a]

Procedure	Time (min)	Volume (mL)	Volume/V_t	Applied or recovered ovalbumin (mg)
Sample application[b]	414	263	57.	53
Desorption[c]	15	9.5	3.3	47

[a]Column: DEAE-Toyopearl 650S (d_p = 40 µm, d_c = 0.9 cm, Z = 4.5 cm, V_t = 2.9 mL). Flow rate: 0.63 mL/min. Initial buffer: 14 mM Tris-HCl containing 0.03 M NaCl (pH 7.7). Sample was 0.02% ovalbumin dissolved into the initial buffer. The desorption was performed with 14 mM Tris-HCl buffer containing 0.3 M NaCl (pH 7.7). Temperature: 25°C. The ovalbumin concentration was determined by the absorbance at 280 nm.
[b]Since the sample was applied until the column was saturated with the sample, that is, the protein concentration at the outlet of the column = the initial concentration, about 11% of the total amount of ovalbumin applied to the column was washed out. In other words, 47 mg ovalbumin was retained in the column before desorption.
[c]The volume of the fraction recovered was approximately 3 mL. A nearly 70-fold concentration was obtained.

resolution of a sample containing a salt is better than that without the salt. He considered that when the sample containing the salt is applied to the column, a decreasing salt gradient is formed in the column, which is responsible for the improvement in resolution.

8.6 SUMMARY AND ADDITIONAL COMMENTS

In Sec. 8.1, the model proposed in Sec. 2.2.2 was experimentally verified. Good agreements between theoretically calculated and experimental curves, as shown in Sec. 8.1, indicate the accuracy and wide applicability of that model. A similar approach

was employed by Pitt (1976) for the investigation of the effects
of various parameters on the elution behavior of small molecules
in gradient elution. His model is based on the mass balance
model in which a basic equation is derived from Eq. (2.24) in
Sec. 2.1. The linear gradient is considered ideally established.
Numerical solutions are obtained by a finite difference technique
with the experimentally determined distribution coefficient as a
function of a buffer concentration. The peak positions are also
calculated by a simplified method proposed by Drake (1955) and
by Freiling (1955). It is quite interesting to compare Pitt's computed results with the results given in Sec. 8.1, since the
differences between the elution behavior of proteins (high-molecular-weight substances) and that of small molecules may be
clarified. One of the most remarkable differences is that the
zone sharpening effect is very weak in his results. This is
attributed to the weak dependence of the distribution coefficient
on the elution buffer concentration. Therefore, the peak width
can be predicted by Eq. (2.87) within small deviations. Equation
(2.87) is based on the assumption that the peak width in
gradient elution is equal to that in isocratic elution, with the
elution buffer of the concentration corresponding to that at the
peak position in gradient elution. Jandera and Churacek
(1974a, b) have reported that the agreement is adequate between
the results calculated by this simplified equation and the experimental results for small molecules. In contrast, for the
prediction of the peak width of proteins in gradient elution, it
is necessary to include the zone sharpening effect, as discussed
in Sec. 8.2.

The model given in Sec. 2.2.2 requires the data for adsorption equilibria and the number of plates. It was found that the

adsorption equilibria between proteins and ion exchangers play
important roles in both stepwise and gradient elution. Although
adsorption equilibria should be known as functions of both the
protein concentration and the ionic strength, they cannot be determined theoretically but must be measured experimentally.
Several workers have investigated adsorption equilibria and reported qualitative theories (see Chap. 3). Theoretical investigation of adsorption equilibria would enable us to predict elution
behavior more accurately.

The number of plates is another important parameter in this
model. It was found that the effect of the number of plates
for salts N_p' was negligible except for stepwise elution with
high-ionic-strength buffers. On the other hand, the number of
plates for proteins plays a significant role in zone spreading,
and in most cases the peak width σ_θ is inversely proportional
to the square root of N_p. The method of determining the number of plates described in Sec. 2.2.2 is successful because
proteins are subjected to an ionic strength of narrow range
while traveling down through the column, since the distribution
coefficient is strongly dependent on the ionic strength. In
other words, the distance that the protein zone moves, governed
by the ionic strength near I_R, is quite long compared with that
governed by an ionic strength lower than I_R, as shown in Fig.
2.13 of Sec. 2.2.1. However, there are some cases for which
the applicability of this method is not confirmed at present.
One is the case in which the distribution coefficient is not as
strongly dependent on ionic strength as that in the study under
discussion here. Another is the case in which the initial concentration of the protein is so high that even the concentration
at the outlet does not obey the linear isotherm shown in Fig.

8.7. In this figure, the same N_p values are employed for all
the elution curves. Actually, the calculated elution curves were
not very different from the experimental curves. However,
further investigations are needed. In addition, since the model
presented in Sec. 2.2.2 is based on the plate theory, we cannot
apply it to the case in which the effect of mass transport between the mobile and the stationary phases is significant (asymmetric elution curves due to a low gel-phase diffusion coefficient, as shown in Fig. 2.7). As already mentioned, the elution curves of ovalbumin on DEAE-Sephadex A-25 seem to belong
to this case. However, these conditions must be avoided to obtain a good separation and therefore it is not a serious limitation
to this model.

A simple method for predicting ionic strength at the peak
position I_R presented in Sec. 2.2.3 was experimentally verified
in Sec. 8.2. This method is based on the simple equilibrium
model. As stated previously, the analytical solutions for the
peak retention time t_R can be obtained when the distribution coefficient K of proteins is assumed to have an exponential or a
power-law dependence on the salt concentration (ionic strength)
(Drake, 1955; Morris and Morris, 1964; Jandera and Churacek,
1974a). Solutions when K exponentially depends on I were
employed in order to predict t_R in the linear gradient elution
experiment from K obtained by the isocratic elution or batch experiment for reversed-phase liquid chromatography (for example,
Jandera and Churacek, 1974b; Stadalius et al., 1984), for the
IEC of proteins (Stout et al., 1986), and for hydrophobic interaction chromatography (Chang et al., 1980). Parente and Wetlaufer (1986) employed the solution when K has a power-law
dependence on I for investigation of the relation between the

retention time in isocratic elution and that in linear gradient
elution. This approximation was also used by Gibbs and Lightfoot (1986), who presented a simple equation that describes the
elution curve. The power-law dependence is also used in Sec.
8.2.

The correlation shown in Fig. 8.13 was found to be successful for the prediction of the peak width in linear gradient elution. This correlation was based on the asymptotic solution derived from a quasi-steady-state model [Eq. (2.86)]. Sluyterman
and his colleagues (Sluyterman and Elgersma, 1978; Sluyterman and
Wijdenes, 1978) also employed a quasi-steady-state model for
analyzing chromatofocusing. However, their results were not
directly applicable to the prediction of the elution curve owing
to uncertainty about the value of the dispersion coefficient and
the range in which the assumption of a quasi-steady state does
not lead to serious errors. Snyder and his colleagues (1979,
1983) developed a model for predicting the peak width in linear
gradient elution reversed-phase chromatography. This model
assumes that K exponentially depends on I. Since only two
constants are involved in the K versus I relation, at least two
experiments can give the values of these parameters. The peak
width is related to the (1) K value when the zone is the mid-
point of the column, and to (2) the parameters describing the
contribution of the zone sharpening effect and the (anomalous)
zone spreading effect (Stout et al., 1986). This model predicted
the experimental results of the IEC of proteins fairly well (Stout
et al., 1986) and was further elaborated by Stadalius (1987).

It is very important to know how the resolution R_s varies
with the column and operating variables. By using the equation that relates R_s and the variables, we can easily optimize

the separation with a small amount of experimental data. The
experimental as well as theoretical results shown in Sec. 8.3 tell
us several interesting and important findings on the R_s in the
linear gradient elution. In general, R_s can be increased with a
decrease in the slope of the gradient, a decrease in the flow
rate, and a reduction in the particle diameter. Snyder et al.
(1983) presented an approximate equation that relates R_s to the
variables. This equation was tested against the experimental
conditions (Stout et al., 1986). It is interesting to note that
this equation is reduced to Eq. (8.12), although the theoretical
approaches are different. Although the column length Z can be
low under the usual conditions, there is an upper limit of R_s ob-
tainable for short columns, as shown in Fig. 8.33. Thus when
the higher resolution is desired, Z must be increased. It is also
noteworthy that in the range where Eq. (8.12) is valid (see Fig.
8.33), R_s at a constant gradient time t_G is dependent on neither
the flow rate nor Z (Snyder et al., 1983). This was verified
experimentally in previous reports (Kato et al., 1982a, 1983a,
1985a; Stout et al., 1986).

It is expected that a protein is adsorbed to anion exchangers
above its isoelectric point pI and to cation exchangers below the
pI. In fact, the choice of ion exchanger is usually made on
the basis of this concept. However, this so-called net charge
concept seems to be an oversimplification of the actual adsorption equilibria between the protein and the ion exchanger, as
pointed out by Kopaciewicz et al. (1983) and by Haff et al.
(1983) (see also Regnier, 1984; Richey, 1984). Kopaciewicz et
al. (1983) carried out linear gradient elution experiments with
the same slope of the salt gradient at different pH values. The
peak retention time t_R was then plotted against the pH. If the

net charge concept holds, the resulting plot should show the minimum t_R, which is determined only by the distribution coefficient owing to size exclusion (gel filtration mode) at the pI for both anion and cation IEC columns. Furthermore, t_R increases with the pH for the anion IEC column and with the decrease in pH for the cation IEC column; the increase in t_R corresponds to the increase in net charge of the protein (see Fig. 8.47). Based on this consideration, Kopaciewicz et al. (1983) made a plot of t_R versus pH, which they call a retention map, for various proteins. They found that none of the proteins employed in their study showed such a simple relation as shown in Fig. 8.47, and many proteins adsorbed to both the anion and cation IEC columns even at their pI. Although many factors may contribute to this phenomenon, such as hydrogen bonding, conformational change of the proteins, and a double electrical layer, they pointed out the importance of the charge asymmetry of proteins. If the net charge concept is valid, proteins with the same pI cannot be separated by IEC. However, as pointed out by Kopaciewicz et al. (1983), the experimental results by Brautigan et al. (1978), in which isomers of substituted cytochromes C having the same net charge were separated by IEC, support the contribution of charge asymmetry to the adsorption mechanism. Kopaciewicz et al. (1983) further suggested that a protein may orient itself so that the charges of local regions become opposite to those of ion exchangers. A similar conclusion was reported by Haff et al. (1983). They also made a retention map similar to that of Kopaciewicz et al. (1983) and compared the results with those of the electrophoretic titration experiment. In general, the retention time increased with the net charge estimated from the electrophoretic titration experi-

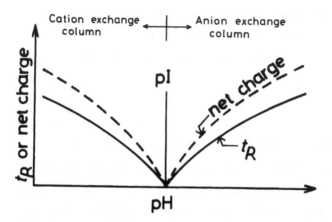

Fig. 8.47 Graphic representation of the peak position and net
charge of a protein as a function of pH. t_R = peak position
(peak retention time), pI = isoelectric point of a protein. Although t_R never reaches 0, it is written as 0 at the pI for the
sake of simplicity. (From Kopaciewicz et al., 1983.)

ment. However, the elution behavior of several proteins could
not be expected from the electrophoretic titration experiment.
Haff et al. (1983) suggested that this behavior was attributed
to charge asymmetry and charge heterogeneity. Hydrophobic
interaction between proteins and the stationary phase may also
contribute to the retention of proteins on IEC columns as well
as electrostatic interaction (Hofstee, 1976; Kopaciewicz et al.,
1985). The pI vs. t_R plot was prepared by Kadoya et al. (1985).

The plot of GH versus I_R shown in Sec. 8.2 can also serve
as a method for determining the nature of a protein and for
surveying the optimum chromatographic conditions, since the
plot implicitly reflects the adsorption equilibria between the
protein and the IEC column, as described in Sec. 8.2. Yamamoto et al. (1986b, 1987d) made these GH versus I_R plots for
various standard proteins with both anion and cation IEC columns

(Fig. 8.48). As shown in Fig. 8.48, as the pH approaches

the respective pI of the protein, the curve of GH versus I_R

shifts to lower I values and also its slope becomes shallow. This

corresponds to the relations extracted from Eq. (8.6). How-

ever, near the pI, the elution behavior of the protein changed

markedly with a small change in pH. For example, both β-

lactoglobulins A (LG-A) and B (LG-B) are adsorbed on either

of the anion or cation IEC columns at pH 5.2. This is in agree-

ment with the results of Haff et al. (1983) and of Kopaciewicz

et al. (1983). Ovalbumin (OA) was not retained on the cation

IEC column at pH 5.2. These results also show that the reten-

tion mechanism of proteins on IEC columns is not explained by a

simple "net charge concept." Therefore, we must rely on the

experimental methods, such as the GH versus I_R plot or the

retention map, for determining the chromatographic conditions.

Let examine the optimum chromatographic conditions from Fig.

8.48. To achieve a fine separation, the values of I_R of proteins

at a certain GH should differ considerably. LG-A and LG-B are

not resolved but are eluted as a single peak on the CM IEC

column. Resolution of LG-A and LG-B is good at pH 5.2 on

the DEAE IEC column, although resolution of OA and LG-B is

poor. The best resolution is obtained at pH 5.6 on the DEAE

IEC column (see Fig. 8.46). However, when both the CM and

DEAE IEC columns are employed, a more efficient separation can

be performed at pH 5.2, as shown in Sec. 8.5 (see Fig. 8.45).

It is advisable to examine the effect of pH on resolution since

resolution is often high at a particular pH, as shown earlier

(Haff et al., 1983; Richey, 1984).

It is often observed that some proteins show anomalous peak

broadening as pointed out previously (Stout et al., 1986; Ghrist

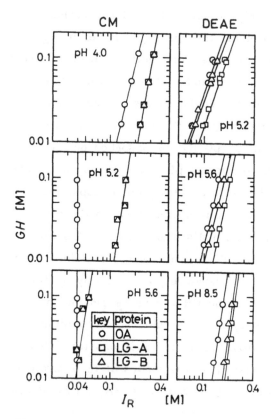

Fig. 8.48 GH versus I_R for ovalbumin (OA), β-lactoglobulins
A (LG-A) and B (LG-B) with DEAE and CM MPIEC columns at
various pH values. I_R = ionic strength [= the molarity of NaCl
+ that of the buffer (0.01)] at the peak position (see Fig.
2.17). The linear gradient elution was performed by increasing
the NaCl concentration from 0.03 M at a fixed pH with DEAE-
or CM-Toyopearl 650S MPIEC columns (d_p = 40 μm, d_c = 1.6
cm, Z = 15 cm). The isoelectric points (pI) of OA, LG-A, and
LG-B are 4.7, 5.1 and 5.2−5.3, respectively. (Righetti and
Caravaggio, 1976; Righetti et al., 1981.) From Yamamoto et al.
(1987d). Reproduced by permission of the American Institute
of Chemical Engineers.

et al., 1987). Microheterogeneity of proteins is a possible reason for this phenomenon. For example, albumin was fractionated
to different components with respect to free sulfhydryl contents
by DEAE IEC (Hartley et al., 1962; Janatova et al., 1968). The
microheterogeneity of ovalbumin due to the differences in car-
bohydrate composition and phosphate content was also found by
the HPIEC method (Peterson and Torres, 1983; Vanecek and
Regnier, 1980). When the resolution power of IEC columns is
not sufficient, a number of small peaks will be included in the
large peaks. This will result in a large but wide elution curve
(Regnier, 1983). Another possible cause of this behavior is the
conformational change of proteins due to their denaturation.

It is well known that an improvement in resolution is always
accompanied by such disadvantages as an increase in the separation time and the pressure drop (mechanical energy) and a
decrease in the peak concentration. Therefore, the choice of
the experimental conditions must be made case by case by
chromatographers. The choice of the elution method, that is,
gradient or stepwise elution, is also strongly dependent on the
respective purposes. If the rapid separation of large amounts
of sample and/or the concentrated fraction is needed, a well-
designed stepwise elution may be more suitable than gradient
elution. In biospecific affinity chromatography (AFC), only a
desired protein is adsorbed to the AFC column and subsequent
one-step desorption yields a highly purified fraction of the
protein. Thus measurement of the adsorption characteristics of
the desired protein gives sufficient information on the design
calculation procedure (Arnold et al., 1985b, c; Arnold and
Blanch, 1986; Chase, 1984a, b; Katoh and Sada, 1980; Sada
et al., 1984, 1985). In contrast a number of contaminant pro-

teins included in the sample as well as the desired protein are
adsorbed on the IEC column during the initial stage. These
contaminants must be eliminated during the subsequent elution
process. Thus information about the contaminant proteins that
are eluted near the desired protein peak is also needed. This
makes the design calculation procedure of IEC of proteins extremely difficult. In any case, we recommend that linear gradient elution experiments with several different slopes of gradients
be performed as preliminary experiments since the results can
serve as fundamental data for the rapid determination of the
optimum chromatographic conditions for both linear and stepwise
elution, as mentioned earlier in this chapter. Several examples
of such procedures are shown in Chap. 10.

It is also noted that the models and methods described in
this chapter are applicable to other chromatographic methods
using different separation modes. For example, a method for
predicting the peak position given in Sec. 8.2 and a correlation
between R_S and the operating variables described in Sec. 8.3
were successfully applied to the hydrophobic interaction chromatography of proteins (Yamamoto et al., 1987c, d).

9

Large-Scale Operation

As mentioned in the introductory chapter, the development of biotechnology requires large industrial-scale liquid chromatography (LC), also called process LC, for finely purifying a large amount of bioproducts. The large preparative-scale LC is also useful for laboratory researchers who wish to obtain a substance of interest that is present in very low concentrations in raw materials.

This chapter deals with large-scale ion-exchange chromatography (IEC)—general considerations (Sec. 9.1), applications (Sec. 9.2), and several problems with the maintenance of the ion-exchange chromatographic column (Sec. 9.3). A good review of the large-scale LC of proteins is given by Janson and Hedman (1982). Recent advances in large-scale LC are found in the literature (for example, Vol. 363 of the *Journal of Chromatography*, 1986; Dwyer, 1984). Reviews on methods for the large (industrial) scale purification or recovery of bioproducts, including LC, are also informative (Bailey and Ollis, 1986; Charm and Matteo, 1971; Dunnill and Lilly, 1972; Porath, 1972;

Verrall, 1985; LeRoith et al., 1985; Vol. 3, No. 1 of Biotechnology Progress, 1987).

9.1 GENERAL CONSIDERATIONS

It is desirable that the separation of proteins by IEC be scaled up so easily on the basis of the data obtained with a small column. Actually, there are several difficulties in the scaling up of IEC:

1. Mechanical stability of packing materials (gels)
2. Design of the column tubes and the end fittings
3. Packing of large columns
4. Prediction of column performance from the data obtained with small columns
5. Column dimension (height-diameter ratio Z/d_c and column-particle diameter ratio d_c/d_p)

9.1.1 Mechanical Stability of Packing Materials (Gels)

A relation between the pressure drop and the flow rate for soft compressible gels is not linear, as was shown in Chap. 7. Furthermore, the maximum flow rate usually decreases with an increase in the column diameter and/or the column length, as shown schematically in Fig. 9.1 (Janson, 1971; Porath, 1972; Janson and Hedman, 1982).

Several special devices for supporting compressible gels have been presented to avoid bed compression in large columns. Such devices include mixing soft gels with inert rigid particles or the insertion of rods or concentric tubes inside the column (Sada et al., 1982). However, these devices not only lower the column efficiency but also make packing of the gel into the column tube difficult. A solution to this problem is to stack short, large diameter columns. The utility of such stacked columns for

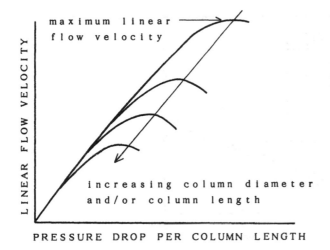

Fig. 9.1 Schematic relation between pressure drop and flow rate for soft gel columns.

soft gel has been claimed (Porath, 1972; Katoh and Sata, 1986b; Janson and Hedman, 1982).

Another approach to obtain an efficient large column is to use such rigid ion exchangers as those employed in high-(HPIEC) or medium-performance IEC (MPIEC), although the column design must be such that the column is resistant to high-pressure forces (see Sec. 9.1.2). Figure 9.2 shows the dependence of the pressure drop per unit column length on column dimension for medium-performance gel filtration chromatography packed with 44 μm particles (Yamamoto et al., 1986a). A linear relation is found between the pressure drop per unit length and the linear mobile velocity up to high flow rates irrespective to the column dimension. Furthermore, this linear relation can be predicted by the Kozeny-Carman equation (7.1), which means the gel bed is neither deformed nor compressed.

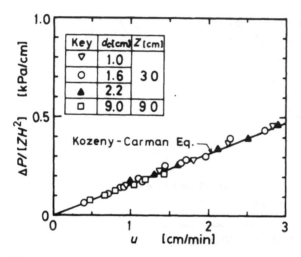

Fig. 9.2 Pressure drop per unit column length versus linear mobile-phase velocity measured for medium-performance gel filtration columns of various column dimensions. H is $(1 - \varepsilon)/\varepsilon$, where ε = void fraction. The values of ε ranged between 0.33 and 0.38. ΔP = total pressure drop for gel beds measured at 20°C. Z = column length, d_c = column diameter, and u is the linear mobile-phase velocity. The packing material (gel) employed is Toyopearl HW55F (d_p = 44 μm). The line was calculated by Eq. (7.1). (From Yamamoto et al., 1986a.)

The fraction of gel beads contacting the column wall in a small-diameter column is larger than that in a large-diameter column. Accordingly, the gel beads are prevented from deforming owing to high pressures in a narrow column. Therefore, we should keep in mind the compression of the gel bed in the large-diameter column even when the gel is rigid.

9.1.2 Design of Column Tubes and End Fittings

Materials for the column tube and the end fittings must be resistant to pressure, solvents, and corrosion. When they are transparent, we can observe movement of the sample zone and

fouling of the gel bed. It is desirable that the column be autoclavable. A detailed discussion of column material is given by
Janson and Hedman (1982). As they stated, for low-pressure
medium-performance columns of relatively small volumes (d_C =
4—10 cm and Z = 10—30 cm), glass or plastic such as polymethylpentene and polycarbonate, may be employed. For large scale
and/or high-performance IEC, stainless steel columns must be
used. In the IEC of proteins, aqueous solutions containing
salts are commonly employed. Thus we must pay attention to
corrosion in the stainless steel column. Although an organic
solvent is seldom used, the IEC column is periodically washed
with acid and alkaline solutions (see Secs. 6.6 and 9.3). Glass
and some plastics are not resistant to strong alkaline solutions.

As already stated, the column length for IEC can be reduced
considerably compared with that for gel filtration chromatography
(GFC), since the resolution can be compensated for by adjustment of other parameters. Therefore, a flat thin column of the
low height-diameter ratio Z/d_C is preferred for large-scale
separation. However, the design of larger diameter columns
must be made carefully since the homogeneous distribution of
liquid onto the whole surface of the gel bed becomes difficult
with the increase in column diameter. In addition, several
problems, such as a device for sample introduction, type of
pump, and the material of the column tube, may be accompanied
by an increase in column dimensions.

Figure 9.3 shows the design of a large commercial column
tube that has six inlet ports and a fine mesh (10 μm) polyamide
fabric and coarse mesh (0.93 × 0.61 mm) support net (Janson
and Hedman, 1982). This column (d_C = 37 cm and Z = 15 cm)
was originally designed as a stacked column with soft gels. The

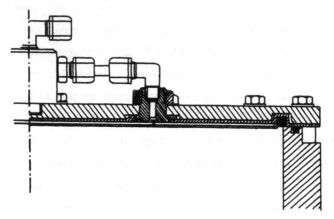

Fig. 9.3 Half-cross section of the end fitting of a large commercial column (d_c = 37 cm, Z = 15 cm). (Reproduced from Janson and Hedman, 1982, by permission of the authors and Springer-Verlag.)

column material is polymethylpentene, which is transparent (Pharmacia Fine Chemicals). The height equivalent to a theoretical plate (HETP) as well as the elution curve for this column, which is packed with gel filtration media, is almost the same as for a small column (5 cm in diameter and 15 cm in length) (Janson and Hedman, 1982). Janson (1971) reported that the zone spreading in the column end fittings can be reduced and a homogeneous flow through the whole gel bed cross section can be achieved by increasing the number of inlet ports. This is because the time needed for the inlet (or outlet) flow to cover the whole gel bed surface is reduced with the increased number of inlet (or outlet) ports.

On the other hand, Nishimoto et al. (1987) found that the column efficiency with a single port is better than that with three ports (Fig. 9.4). Their column design is shown in Fig. 9.5, which is now commercially available. The superficial velocity

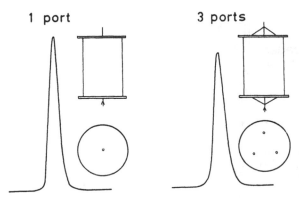

Fig. 9.4 Effect of the number of the inlet (and outlet) ports on zone spreading. The elution curves of 5% acetone pulses (184 mL) at a superficial velocity of 1.6 cm/min were obtained on DEAE-Toyopearl 650M IEC columns (d_c = 31 cm, Z = 40 cm). (Reproduced from Nishimoto et al., 1987, by permission of the authors.)

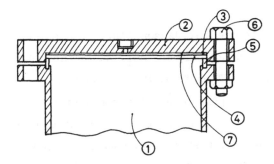

Fig. 9.5 Cross section of the end fitting of a large commercial column made of stanless steel: (1) packing materials, (2) flange (SUS316), (3) seal plate (PTFE), (4) filter (SUS316L), (5) seal ring (PTFE), (6) bolt and nut (SUS316), (7) space. The column dimensions are d_c = 10.8 and Z = 30 cm, d_c = 30 cm and Z = 40 cm, or d_c = 60 cm and Z = 40 cm. (Reproduced from Nishimoto et al., 1987, by permission of the authors.)

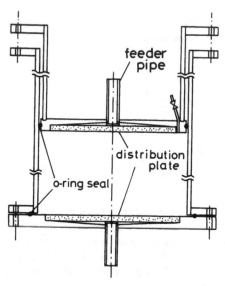

Fig. 9.6 Vertical section of a large column. Liquid flow is distributed radially through six openings 1 mm in diameter on the wall of a dead-end feeder pipe, which interrupts the axial momentum of liquid flow. A sintered glass plate is mounted in a tapered end plate fixed to the feeder pipe as a distribution plate. Column diameter is 10 or 20 cm. (Reproduced from Sada et al., 1987, by permission of the authors and the publisher.)

is much higher than that employed by Janson (1971). A sintered metal plate filter is used not only for support of the gel bed but also as a distributor. Therefore, the pressure drop through the column and the support result in a homogeneous distribution of the flow onto the gel bed surface. A similar conclusion was reached by Sada et al., (1987). They measured the HETP values of NaCl with soft gels (Sepharose 4B) and rigid silica gels packed in the column tube shown in Fig. 9.6. The HETP values for the silica gel was higher than those for the soft gel (Fig. 9.7). The pressure drop through the silica gel column was much lower than that through the soft gel column.

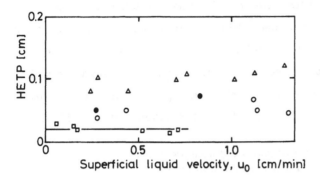

Fig. 9.7 Relation between HETP and superficial velocity u_0.
Sample: 2 M NaCl. Packing materials used are nonspherical
silica (d_p = 63–200 μm) ($\triangle$, $\circ$, $\bullet$) or Sepharose 4B (d_p = 60–
140 μm ($\square$). ($\square$) d_c = 10 cm, Z = 15.5 cm, and No. 2 sintered
glass; ($\triangle$) d_c = 10 cm, Z = 14.7 cm, and No. 2 sintered glass;
($\circ$) d_c = 10 cm, Z = 14.9 cm, and No. 4 sintered glass; ($\bullet$) d_c
= 20 cm, Z = 4.7 cm, and No. 4 sintered glass. The average
pore size of the sintered glass is 40–50 μm for No. 2 and 5–10
μm for No. 4. The thickness of the sintered glass is 0.7 cm.
(Reproduced from Sada et al., 1987, by permission of the
authors and the publisher.)

When the pore size of the support filter (sintered glass) was

decreased from 40–50 to 5–10 μm, the HETP values for the

silica gel improved, as shown in Fig. 9.7. They concluded that

the pressure drop through the gel bed and the sintered glass

is responsible for the uniform distribution of flow onto the whole

gel bed surface.

As demonstrated in Sec. 9.2, the large-diameter column with

a single port shown in Figs. 9.5 and 9.6 is adequate for large-

scale medium- and high-performance IEC.

9.1.3 Packing the Column

HPIEC columns are usually packed at much higher pressure than

in usual operation. Therefore, packing large HPIEC columns

requires a high pressure and a high-flow-rate pump. Packing

large HPIEC and MPIEC columns is thus carried out by the
manufacturer. We can pack MPIEC and LPIEC columns of moder-
ate sizes (d_c = 2−10 cm and Z = 10−30 cm) according to the
methods described in Sec. 6.3. An interesting chromatographic
system has been reported in which the gel bed is packed with
axial compression using a mobile piston (Clonis et al., 1986).

9.1.4 Prediction of Column Performance from Data Obtained with Small Columns

Even when packing materials are so rigid that they are not deformed in the column and the column tube and the end fittings
are properly designed, as mentioned, column efficiently may be
dependent on the column dimensions, especially on column diameter, owing to the distribution of mobile-phase velocity in the
column, the so-called wall effect.

It is thought that there is a region near the wall of the
column where the packing density is low and consequently the
flow is faster than that near the center axis of the column.
When the ratio of such a region to the total cross-sectional area
is not negligible, the column efficiency depends on the column
diameter d_c. The wall effect is therefore determined by the
column-particle diameter ratio d_c/d_p. Figure 9.8 shows the
relation between HETP and linear mobile-phase velocity u with
the same MPGFC columns as those in Fig. 9.2. The HETP can
be expressed by a single straight line within small experimental
errors regardless of the column dimension. This relation is
expressed by the HETP equation (2.46). In order to investigate
the wall effect, the sample was injected at the center of the top
of the gel bed, which was 1.0 cm in diameter and 30 cm long.
The zone was not observed until it reached the end of the

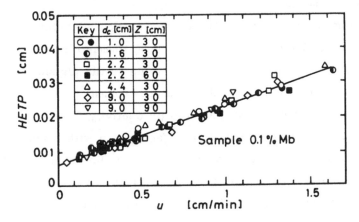

Fig. 9.8 Relation between HETP and linear mobile-phase velocity u for medium-performance gel filtration columns of various column dimensions. The sample is 0.1% myoglobin. The sample volume was chosen so that it was around 0.5—1.0% of the total column volume. Experiments were carried out at 20°C with the buffer (pH 7.9) containing 0.3 M NaCl. The gel employed is Toyopearl HW55F (d_p = 44 μm). (•) indicate the HETP values obtained with the central injection method. (From Yamamoto et al., 1986a.)

column, in contrast to the usual sample application onto the whole surface of the top of the column, which gave a horizontal zone. Although very small amounts of tailing and/or leading of the elution curve were eliminated by the central injection method, the HETP itself was not appreciably improved. The wall effect is thought to diminish with an increase in the column diameter. In other words, when the column-particle diameter ratio d_c/d_p is above 200, the column efficiency is not markedly varied. This has been shown previously (Kelly and Billmeyer, 1969; Klawtier et al., 1982; Kaminski et al., 1982). It is safe to ignore the wall effect in the IEC of proteins unless a narrow column and a large particle are employed.

9.1.5 Column Dimensions

The column dimensions, such as the column-particle diameter ratio $m = d_c/d_p$ and the column length-column diameter ratio $n = Z/d_c$ must be changed in scaling up owing to the physical properties of the ion-exchange gel and/or the column designs already mentioned. Let us calculate the productivity of two columns that are equal in volume but different in n with simplified assumptions. The superficial velocity u_0 (or linear mobile-phase velocity u) is taken to be the same (column 1: $d_{c,1}$, Z_1, V_{t1}, n_1; column 2: $d_{c,2}$, Z_2, V_{t2}, n_2; note that $V_{t,1} = V_{t,2} = V_t$).

When the elution is performed by one-step desorption (type I elution described in Sec. 8.4), the ratio of the separation time Z/u_0 of the two columns becomes (Janson and Hedman, 1982)

$$\frac{Z_2/u_0}{Z_1/u_0} = \left(\frac{n_2}{n_1}\right)^{2/3}$$

since $V_t = (\pi/4)nd_c^3 = (\pi/4)Z^3/n^2$. Therefore, the productivity increases with decreasing $n = Z/d_c$.

In addition to the preceding assumptions, we assume that the $Z/GHu = Z/ug(V_t - V_0)$ values of the two columns are set to be equal in linear gradient elution. Both the separation time t_R and the R_s values are thus the same, as is clear from Eqs. (2.81) and (2.84) and the discussions given in Sec. 8.3.9. That is, n is independent of the productivity.

The $n = Z/d_c$ value of analytical IEC columns is around 10. For large columns, n is smaller than 2 and sometimes below 1.0. In general, it is advantageous to increase the column diameter d_c in the scale-up of IEC. However, it seems difficult

to accomplish a homogeneous distribution of flow into a short
column of very large diameter, for example, 40 cm in diameter
and 5 cm high, although the height of analytical HPIEC columns
is 5—8 cm.

When the particle diameter d_p is changed, the other parameter should be adjusted to obtain a desired resolution in linear
gradient elution according to Eq. (8.12). In stepwise elution,
when d_p is increased a decrease in flow rate and/or the increase in column length is needed to obtain the same separation
efficiency.

9.2 APPLICATIONS

Unfortunately, only a few reports on the application of large-
scale IEC to protein separation are found in the literature in
which the details of the chromatographic conditions are described.
We explain only the following four typical applications: linear
gradient elution with a low-pressure IEC column (Sec. 9.2.1),
stepwise elution with cross-linked agarose ion-exchange gel
columns (Sec. 9.2.2), linear gradient elution with medium- and
high-performance IEC columns (Sec. 9.2.3), and stepwise elution
with a medium-performance IEC column (Sec. 9.2.4).

9.2.1 Linear Gradient Elution with a DEAE-
Sephadex A-50 Column

Bruton et al. (1975) performed linear gradient elution for the
separation of methionyl-tRNA synthetase and tyrosil-tRNA
synthetase from *Escherichia coli* with a 75 L DEAE-Sephadex
A-50 column (d_c = 37 cm and Z = 70 cm). The column tube
employed is the same as that shown in Fig. 9.3. As described
in Sec. 8.5, the crude extract was first purified by a batch ad-

sorption-desorption method with DEAE-cellulose. The fraction
(50 L containing 1200 g protein) was then applied to the DEAE
IEC column at a flow rate of 2.5–3 L/hr. After the column
was washed with a 30 L starting buffer solution (100 mM potassium phosphate, pH 8), the linear increase in the potassium
phosphate buffer concentration (1×10^{-6} M/mL) was applied to
the column at a flow rate of 1.8 L/hr ($u_0 = 0.028$ cm/min). The
yield of the two enzymes was 74–79%, and the purification factor was 6.6–6.9. The separation time was 140 h (about 5 days).
The sample load was 16 mg protein per mL column.

9.2.2 Stepwise Elution with DEAE- and CM-Sepharose CL-6B Columns

Plasma protein fractionation is one of the most important purification processes in pharmaceutical technology. The conventional process was the precipitation by ethanol followed by centrifugation (Cohn et al., 1946). The chromatographic separation
process using both anion and cation IEC columns are presented
(Curling, 1980; Janson and Hedman, 1982; Strobel, 1982). In
the first step, while human serum albumin (HSA) is adsorbed to
a DEAE column equilibrated at pH 5.2 and ionic strength I =
0.025, immunoglobulin G (IgG) passes through the column (fraction A in Fig. 9.9). Subsequent desorption by the sodium
acetate buffer (pH 4.5 and I = 0.025) yields the HSA fraction
(fraction C in Fig. 9.9). This fraction is adjusted to pH 4.8
and I = 0.07 so that it can be adsorbed to a CM column equilibrated under these conditions. The HSA fraction is then recovered from the CM column (fraction G in Fig. 9.9) by desorption
with the sodium acetate buffer (pH 5.5 and I = 0.11).

The gel bead employed in this method is cross-linked agarose.
The column tube is the same as that shown in Fig. 9.3. Al-

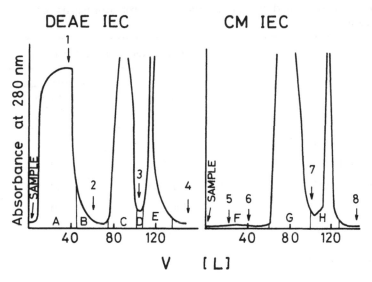

Fig. 9.9 Purification of human serum albumin by large-scale
IEC. Desalted, euglobulin-poor human plasma (36.6 L containing approximately 500 g albumin) was first applied to a DEAE-
Sepharose CL-6B column (d_c = 37 cm, Z = 15 cm, V_t = 16 L)
at room temperature. u_0 = 0.37 cm/min. A = IgG fraction;
B, D, E = discard fractions; C = albumin fraction (23.5 L containing approximately 500 g albumin). After pH and ionic
strength I adjustment to pH 4.8 and I = 0.07, the albumin fraction C was applied to a CM-Sepharose CL-6B column (d_c = 37
cm, Z = 15 cm, V_t = 16 L) at room temperature. u_0 = 0.33
cm/min. F, H = discard fractions; G = albumin fraction (36.6
L containing approximately 550 g albumin). The arrows in the
figure show the change in the elution buffer (see Table 9.1).
(Reproduced after slight modification from Janson and Hedman,
1982, by permission of the authors and Springer-Verlag.)

though the particle diameter is not described, the average par-

ticle diameter is probably around 100 μm (see Chap. 7). Since

this bead is more rigid than the soft dextran ion-exchang gel

mentioned DEAE-Sephadex A-50, the flow rate is higher (u_0 =

0.33 ~ 0.37 cm/min). The bead is not resistant to the pressure up

to several atmospheres, however (Larre and Gueguen, 1986). The

Table 9.1 Buffers Used in Ion-Exchange Chromatography of
Human Plasma[a]

		Sodium acetate buffers	
Column[b]	pH	Ionic strength	Approximate volume (L)
DEAE-Sepharose CL-6B			
1. First elution	5.2	0.025	20
2. Second elution	4.5	0.025	30
3. Final elution	4.0	0.15	40
4. Regeneration	5.2	0.025	40
CM-Sepharose CL-6B			
5. First elution	4.8	0.07	20
6. Second elution	5.5	0.11	60
7. Final elution	8.1	0.4	40
8. Regeneration	4.8	0.07	40

Total volume in DEAE-Sepharose chromatography: 130 L
Total volume in CM-Sepharose chromatography: 160 L

[a]Volumes for one complete cycle on 16 L columns.
[b]Numbers correspond to those in Fig. 9.9
Source: From Janson and Hedman, 1982.

buffer solutions used and the operating conditions are summarized
in Table 9.1. One cycle for this separation requires about 13 h.
The sample load is approximately 30 mg HSA per mL column,
which is much higher than that in analytical IEC (see Sec.
8.3.8).

The yield of this method is 95%, and the purity of HSA is
97%. It is said that this process is now scaled up to 150 L
columns for industrial-scale production. For the same purpose,

Saint-Blancard et al. (1982) employed a 50 L DEAE-Trisacryl IEC column (d_c = 35 cm and Z = 50 cm) and a 50 L blue Trisacryl affinity column (d_c = 35 cm and Z = 50 cm).

The isoelectric point pI of HSA is around 4.8 (Righetti et al., 1981). The pH of the initial buffer is thus not so far from the pI of HSA. This is one of the reasons that HSA is adsorbed to either the anion (DEAE) or cation (CM) IEC columns at a pH near 5.0. It also should be mentioned that the protein content of plasma is about 8%, about 60% of which is HSA. The concentration of the desired protein in the starting material is thus not so low as that in fermentation or cell culture broths (Dwyer, 1984; Bailey and Ollis, 1986).

9.2.3 Linear Gradient Elution with Medium- and High-Performance IEC Columns

In Sec. 9.1, the versatility of medium-performance liquid chromatography, that is, a high column efficiency and a high mechanical stability of gels, was shown. Yamamoto et al. (1987b) have presented a strategy for the large-scale purification of proteins by using both HPLC and MPLC. They performed the large-scale linear gradient elution IEC of crude β-galactosidase on medium-performance ion-exchange chromatography columns. Factors affecting the purificaiton factor, such as the slope of the gradient, the particle diameter, and the sample load, were also investigated (see Fig. 8.26).

As mentioned in Sec. 9.1, the height-diameter ratio Z/d_c and the particle diameter d_p must be changed often in the scale-up. When the slope of he gradient g is adjsuted so that GH = $g(V_t - V_0)$ becomes equal, the peak positions in V/V_0 are similar, as described in Secs. 2.2.3 and 8.2.1. The elution curves thus obtained for 23 mL and 2.7, 30, and 113 L MPIEC columns (d_p = 65 μm) are shown in Fig. 9.10. The design of these columns is shown in Fig. 9.5. When we consider the experimental difficulties in preparing a large volume of buffer and sample solutions, making a large-scale linear increase in NaCl, packing large columns, and determining V_0, we can conclude that the peak positions in V/V_0 as well as the peak shapes are quite similar. A relatively high flow rate (u_0 = 0.6 cm/min or u = 1.5 cm/min) was obtained even for the largest column (d_c = 60 cm, Z = 40 cm, and V_t = 113 L). The sample load is about 3.3 mg crude enzyme per mL column, but it can be increased to 17 mg crude enzyme per mL column with a slight decrease in the purification factor (see Fig. 8.26).

Figure 9.11 shows the elution curves of the same sample as in Fig. 9.10 on high-performance ion-exchange chromatography columns (d_p = 20−40 μm). The sample load is 6.7 mg crude enzyme per mL column (Fig. 9.11a) or 3.3 mg crude enzyme per mL column (Fig. 9.11b). The flow rate (u_0 = 2 cm/min) is about three times as high as that for the large MPIEC columns shown in Fig. 9.10. The resolution of each peak is better than that in Fig. 9.10, although the separation time is about one-third that in Fig. 9.10. The resolution of the peak appearing at the beginning of the elution is markedly affected by increasing the sample load, but the peaks eluted after 60 min are not markedly varied. From these results, we can see that large-scale HPIEC columns give high productivity as well as high resolution and high speed of separation.

9.2.4 Stepwise Elution with Medium-Performance IEC

Figure 9.12 compares the elution curves obtained with small and large MPIEC columns. The sample is the same as that employed

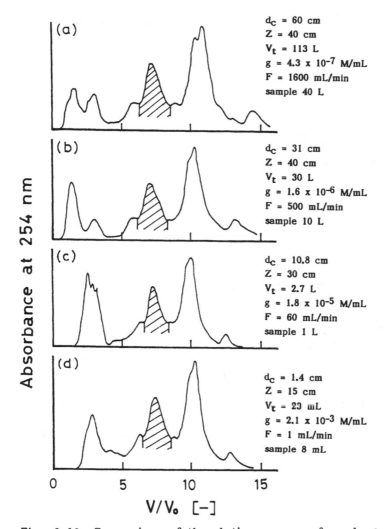

Fig. 9.10 Comparison of the elution curves of crude β-galacto-sidase by linear gradient elution with MPIEC columns of various column dimensions. The MPIEC packing is DEAE-Toyopearl 650M (d_p = 65 μm). The sample (1% crude β-galactosidase) volume is approximately one-third the total column volume V_t. V_0 = column void volume, and V = elution volume from the start of the linear gradient elution. The shaded area indicates the fractions with β-galactosidase activity. The chromatographic conditions are essentially the same as those in Fig. 7.2. The time corresponding to V/V_0 = 15 is 420 min for column a, 370 min for column b, 280 min for column c, and 140 min for the column d. (Reproduced from Yamamoto et al., 1987b, by permission of the authors and Elsevier Publishing Company.)

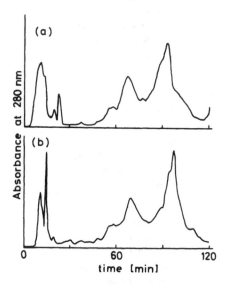

Fig. 9.11 Elution curves of crude β-galactosidase by linear gradient elution with a large-scale HPIEC column. The HPIEC column is TSK DEAE 5PW (d_p = 20–40 μm, d_c = 21 cm, Z = 30 cm, V_t = 10.4 L). The chromatographic conditions are essentially the same as those in Fig. 7.2. g = 3 × 10^{-6} M/mL, GH = 0.0196 M, u_0 = 2 cm/min, u = 5.33 cm/min. The sample (2% crude β-galactosidase) volume is approximately one-third the total column volume V_t = 3.5 L for a and one-sixth V_t = 1.75 L for b. (From Yamamoto et al., 1987f.)

in Figs. 9.10 and 9.11. The details of the method for determining the operating conditions are shown in Sec. 8.4.3. As mentioned in Sec. 8.4.3, the purification factor is lower than that of the linear gradient elution shown in Sec. 9.2.3. However, the sample load (25 mg crude enzyme per mL column) is high and the peak fraction is concentrated. Although the peak posi-

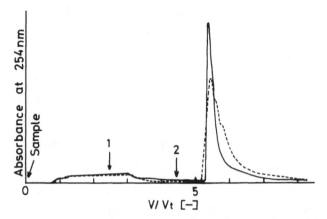

Fig. 9.12 Elution curves of crude β-galactosidase by stepwise elution with a large-scale MPIEC column. The chromatographic conditions are essentially the same as those in Fig. 8.40. The dotted curve represents the elution curve for a small MPIEC column (DEAE-Toyopearl 650S, d_p = 40 μm, d_c = 1.6 cm, Z = 15 cm) shown in Fig. 8.40a. The solid curve shows the elution curve for a large MPIEC column (DEAE-Toyopearl 650M, d_p = 65 μm, d_c = 31 cm, Z = 40 cm, V_t = 30 L). The sample (1% crude β-galactosidase) volume is approximately $2.5 \times V_t$ (73.5 L for the large column and 75 mL for the small column). u_0 = 0.69 cm/min for the large column. The arrows indicate a change in the elution buffer: (1) the initial buffer (washing), (2) the buffer containing 0.15 M NaCl (desorption). The time corresponding to V/V_t = 7 for the large column is 406 min. (From Yamamoto et al., 1987e.)

tions and shapes are similar, the enzyme peak of the large column is higher than that of the small column. This is attributed to the difference in column length. Since the large column is 2.7 times as long as the small column and the linear velocity is the same for the two columns, the number of theoretical plates for the large column is higher than that for the small column. However, the separation time for the large column increases by a factor of 2.7. Similarly, the large columns in Figs. 9.10 and 9.11 are longer than the analytical columns (for

example, Z = 15 cm for the small MPIEC column shown in Fig.
9.10 and Z = 7.5 cm for the small HPIEC column shown in Fig.
7.2). Since the linear mobile-phase velocity u and the parameter GH are kept constant, the throughput (for example, the
fractionated protein/time column volume) of the large column is
lower than that of the small column although the resolution of
the large column is higher. Such a decrease in throughput
with the increase in column volume is due to the upper limit
of flow rate, as mentioned in Sec. 9.1.1. Even when a large
column is packed with rigid particles, such as those used for
analytical HPIEC columns, there is an upper limit of the flow
rate (usually recommended by the manufacturer). The reason
for this is essentially the same as that described in Sec. 9.1.1.
Therefore, in order to increase the throughput, an increase
in sample load or in the slope of the gradient is needed, although both of these cause a decrease in the resolution or the
purification factor.

9.3 PROBLEMS ENCOUNTERED IN ACTUAL OPERATION

The fouling of ion-exchange chromatography columns or packings (gels) is one of the most serious problems in actual operation, as already emphasized. Fouling not only decreases the
column efficiency but also markedly increases the pressure drop
(Gustafsson et al., 1986; Johansson and Stafstrom, 1984; Yamamoto et al., 1986a). It is recommended that the buffer solutions
and samples be filtered through a microfilter (0.22 or 0.45 μm)
before use. However, fouling is unavoidable when the particle
diameter of IEC gels is small and/or a sample contains materials
that are insoluble or are almost irreversibly adsorbed to the ion

exchangers. Even when these conditions are avoided, contaminant is accumulated as the number of uses of the column operation increases. Microbial contamination or bacterial growth is
another type of fouling. These results in an increase in the
pressure drop (Gustafsson et al., 1986; Johansson and Stafstrom,
1984). A good method for preventing the IEC column from
fouling is to insert a small column, called a guard column or a
precolumn, between the inlet of the column and a pump. Packing material for the guard column may be, rigid gel filtration
chromatographic gels, for example, as well as the same packing
material as used in the IEC column. Periodic washing of the IEC
column with appropriate solvents is also an efficient method for
this purpose. Wehr (1984) has reviewed the care of HPLC columns. Curling and Cooney (1982) exemplified how to resolve
problems encountered in large-scale IEC, such as the maintenance
and regeneration of ion exchangers. They recommended the
periodic washing of the IEC column with 0.1 M sodium hydroxide
solution to remove accumulated contaminants. They stated the
following merits in using sodium hydroxide. (1) Its presence in
the final product may be allowed in many cases. (2) It works
as an effective disinfectant. (3) It can solubilize proteins and
lipids, which may serve as substrates for the growth of bacteria
and, therefore, remove them. (4) It is inexpensive and can be
disposed easily. However, some ion exchangers are not stable
in alkaline solutions, such as silica-based packings. In such
cases, weak acidic solutions, such as 1% acetic acid and 1%
phosphoric acid, are used for washing (Wehr, 1984). The use
of formaldehyde (1% v/v) and ethanol (70% v/v) for the preven-
tion of bacterial growth in the column is recommended by Curling
and Cooney (1982). Although organic solvents, such as trifluoro-

acetic aicd, are also effective for washing of the column, they
may deteriorate the column. We should thus follow the washing
procedure recommended by the manufacturer.

High-purity water must be used when the end product
purified by IEC is used for pharmaceutical purposes. In addition, cleaning and disinfection of the system are required in
such cases. Curling and Cooney (1982) reported that the water
used for such systems must be filtered to at least 0.45 μm or
preferably 0.22 μm, must contain a very low concentration of
dissolved gases, and should be pyrogen free. They also discussed a method for the preparation of a large volume of buffer
solution that fulfills these requirements. Ganzi (1984) have also
reviewed methods for preparing high-purity water in the laboratory. According to Curling and Cooney (1982), the ratio of the
cost of water to the total cost of albumin solutions is about 1%.

The life span of IEC columns depends strongly on their
maintenance, including washing, the quality of the buffer solutions, and the properties of a sample. Proper maintenance not
only increases the life span of IEC columns but also gives reproducible results (Johansson and Stafstrom, 1984).

10

Design Calculation Procedure

The separation mechanism of proteins in IEC can be understood
to a great extent through the theoretical considerations and experimental results so far described. However, some readers
may have mathematical difficulties when they try to design IEC
separations according to the theoretical sections of this book.
Therefore, in this chapter, we give several examples of design
calculation procedures to show how the separation is scaled up
and/or the optimum chromatographic conditions are estimated on
the basis of the results obtained from linear gradient elution
experiments with a small column. As already emphasized, the
distribution coefficient $K(I)$ of proteins as a function of ionic
strength I is a very important parameter affecting both the
retention and resolution. Therefore, in this section, we treat
the following two cases:

case 1: the $K(I)$ of proteins to be separated are unknown.
case 2: the $K(I)$ values are known.

In both cases, the particle diameter d_p of the IEC packings is
considered known. In addition, the objective proteins are assumed

to be identified. This can be done, for example, by measurement of the enzymatic activity when the objective proteins are
enzymes. In both linear gradient and stepwise elution experiments, the elution is performed by increasing the salt concentration at a fixed pH. We first treat the design calculation procedure for linear gradient elution.

10.1 LINEAR GRADIENT ELUTION

We focus our attention on the resolution of two components
contained in a sample mixture because this simplifies the design
calculation procedure. Once the column and operating variables
to obtain a desired resolution R_S of the two components are
estimated, the resolution of other components in the sample mixture can also be calculated in a similar manner.

A general flow sheet for the calculation procedure of both
case 1 and case 2 is shown in Fig. 10.1. Since methods for
choosing the ion exchanger (IEC packings) and the buffer solution were discussed in Chaps. 6 and 7, they are omitted in the
flow sheet. As recommended by Regnier (1984), chromatographers should search the literature for the separation similar to
the one they are attempting, since such literature will help them
to determine not only the optimum chromatographic conditions
but also the suitable conditions under which the desired protein
or enzyme is stable. As shown in the figure, we recommend that
the chromatographer first prepare plots of GH vs I_R; the utility
of these plots was stressed in Secs. 8.2 and 8.6. First let us
consider case 1.

Case 1

In order to make the GH versus I_R plot, the distribution coefficient K as a function of ionic strength I must be measured

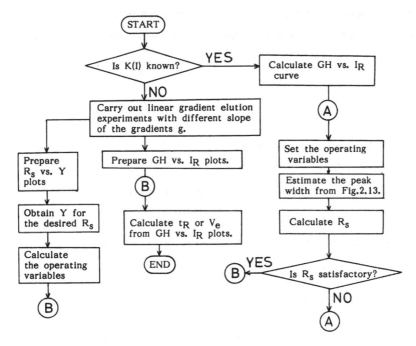

Fig. 10.1 Flow sheet for the design calculation of linear gradient elution.

by a batch (static) experiment, as described in Chap. 3. However, it is rather laborious to carry out such a batch experiment over a wide range of I. Moreover, the batch experiment cannot be employed for measuring K values of a sample containing more than one component unless each component can be detected selectively. Therefore, we recommend that linear gradient elution experiments be carried out in order to prepare the plots of GH versus I_R. In this method, I_R can be measured for several components in a single experiment when their peaks are well resolved. In addition, the choice of the experimental conditions and operation are easy.

Since a small column is sufficient for making the plot, we employ a column of 1.0 cm diameter × 10 cm length. The column volume V_t is then given by

$$V_t = \left(\frac{d_c}{2}\right)^2 \pi Z = A_c Z = 7.85 \text{ mL} \tag{10.1}$$

where d_c = column diameter

   Z = column length

   $A_c = (d_c/2)^2 \pi$ = cross-sectional area of the column

Although the void fraction ε can be determined experimentally, it is assigned as 0.4 to a good approximation. Then,

$$V_0 = V_t \varepsilon = 3.14 \text{ mL} \quad \text{column void volume} \tag{10.2}$$

$$V_t - V_0 = 4.71 \text{ mL} \quad \text{column gel volume} \tag{10.3}$$

As for the pH of the buffer and the type of IEC column, it is good practice to perform a preliminary linear gradient elution experiment with the anion IEC column at a pH near 8.0 since the isoelectric points of many proteins are less than 7.0.

The experimental results given in Sec. 8.2 show that under the usual chromatographic conditions GH must be varied from 0.01 to 0.1 M to obtain a curve of the plot of I_R versus GH. For example, if linear gradients whose slope correspond to GH = 0.1, 0.05, and 0.01 M are to be prepared, the slope of the linear gradient g can be calculated as

$$g = \frac{GH}{V_t - V_0} = \frac{GH}{4.71} \tag{10.4}$$

Then for GH = 0.1 M,

$$g = \frac{0.1}{4.71} = 0.0212 \text{ M/mL} \tag{10.5}$$

Similarly, for GH = 0.05 M, g = 0.0106 M/mL and for GH = 0.01
M, g = 0.00212 M/mL. As a first approximation, the initial ionic
strength I_0 is chosen as 0.03. When a device consisting of two
vessels, shown in Fig. 6.4, is used to make a linear gradient,
g is given by [see Eq. (6.1)]

$$g = \frac{I_f - I_0}{V_G} \qquad (10.6)$$

where I_f = ionic strength of the final buffer

V_G = gradient volume, the sum of the volume of the initial
and final buffer solutions in the vessels at the beginning

If I_f = 0.5 M, V_G to obtain g = 0.0212 M/mL is

$$0.0212 = \frac{0.5 - 0.03}{V_G}$$

$$V_G = 22.2 \text{ mL} \qquad (10.7)$$

If an automatic gradient-making device is available, we
prepare only the initial and final buffer solutions. We should
keep in mind that in many automatic gradient devices we must
set not the gradient volume V_G but the gradient time t_G.
Therefore, for the calculation of V_G in Eq. (10.6), the volumetric flow rate F is needed:

$$V_G = Ft_G \qquad (10.8)$$

When it is not certain whether the proteins contained in the
sample can be adsorbed to the IEC column at the initial conditions, the sample volume should be within 10% of the total column volume. Therefore, we choose 0.5 mL as a sample volume.
The choice of the linear mobile-phase velocity u is dependent on
the particle diameter and mechanical stability of the IEC column

(packing). Unfortunately, only one particle size range is available for most commercial IEC packings for protein separation. Once the IEC packing to be used is chosen, the working range of u is thus fixed. Let us determine the value of u. Let us employ the IEC packing, the particle diameter d_p of which is about 40 μm. As discussed in Sec. 2.1.6, it is desirable to obtain symmetrical peaks such that the dimensionless variable $\phi = (d_p/2)(\bar{D}Z/u)^{-1/2}$ is less than 0.2 in gel filtration chromatography (GFC). In the IEC of proteins, ϕ can be increased up to around 0.5–0.7 since K is larger than 1.0 during elution. The gel-phase diffusion coefficient $\bar{D}$ of proteins is usually 1/5 to 1/10 of the molecular diffusion coefficient D_m, as shown in Fig. 4.2 in Sec. 4.3. We first estimate D_m by assuming that the molecular weight (MW) of a protein is 80,000 (MW of proteins of medium size, such as bovine serum albumin, is around 80,000) and the temperature is 25°C. The viscosity of water at 25°C is 0.89 cP. $D_m\eta/T$ for MW = 80,000 is read from Fig. F.1 in App. F as 2×10^{-9}. D_m is then calculated as

$$D_m = 2 \times 10^{-9}\,\frac{273 + 25}{0.89} = 6.7 \times 10^{-7}\ cm^2/s \qquad (10.9)$$

$\bar{D}$ is thus given by $D_m \times 0.15 = 1.00 \times 10^{-7}\ cm^2/s$ when $\gamma_{sm} = \bar{D}/D_m$ is assumed to be 0.15. By using this $\bar{D}$ value together with d_p and Z, we calculate u as

$$u = \frac{\phi^2 \bar{D} Z}{(d_p/2)^2} = \frac{0.5^2 \times 1.0 \times 10^{-7} \times 10}{(40 \times 10^{-4}/2)^2}$$

$$= 0.063\ cm/s = 3.8\ cm/min \qquad (10.10)$$

Since this u value is the maximum to obtain a symmetrical peak, it is safe to reduce u to achieve a fine separation. If u is set at 3.0 cm/min, the volumetric flow rate F is given by

F = u × cross-sectional area × void fraction

$$= 3.0 \times [(\tfrac{1}{2})^2 \pi] \times 0.4 = 0.94 \text{ mL/min} \tag{10.11}$$

Next, we perform linear gradient elution experiments with different slopes of the gradient as already determined by using this flow rate. The experimental results obtained are interpreted as follows. When the electroconductivity meter is employed, the ionic strength at the peak maximum I_R can be determined experimentally. If this apparatus is not available, I_R is estimated from the peak retention time, t_R according to the approximate procedure.

$$I_R = I_0 + g(Ft_R - V') \tag{10.12}$$

$$V' = V_0 + K'(V_t - V_0) \tag{10.13}$$

where V' = elution volume of a salt employed for increasing ionic strengths. As shown in Table 3.1, the distribution coefficient for salts K' usually ranges between 0.8 and 1.0. As a good approximation, K' is thus taken to be 0.9. V' is thus given as

$$V' = 3.14 + 0.9 \times 4.71 = 7.38 \text{ mL} \tag{10.14}$$

If the observed t_R for a protein contained in the sample at g = 0.00212 M/mL, that is, GH = 0.01 M, is 50 min,

$$I_R = 0.03 + 0.00212 \times (0.94 \times 50 - 7.38)$$

$$= 0.114 \text{ M} \tag{10.15}$$

This value is then plotted against GH = 0.01 M. Similarly, the results with g = 0.0106 M/mL, that is, GH = 0.05 M and g = 0.0212 M/mL, that is, GH = 0.1 M, are plotted in the same graph. If the resolution is not sufficient, we must reduce the flow rate and/or the slope of the gradient. However, the calculation procedure for the preparation of the chart is the same. It is expected from the consideration given in Secs. 8.2 and 8.6 that the curve

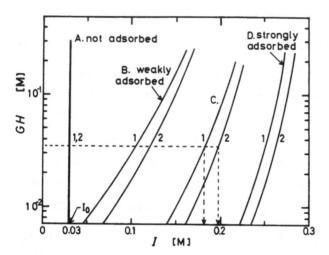

Fig. 10.2 Schematic representation of the relation GH versus I_R. The numbers 1 and 2 imply substances 1 and 2, respectively.

of GH versus I_R changes its shape from that of curve A to that of curve D shown in Fig. 10.2 when the net charge is increased (the increase in pH from pI for anion-exchange columns and the decrease in pH from pI for cation-exchange columns). Moreover, since the shape of the curve, which reflects the nature and characteristics of the proteins, is different from protein to protein, there may be a certain pH at which the distance between the curves of the two proteins is quite long and hence the separation can best be accomplished as discussed in Sec. 8.6. Therefore, it is advantageous to prepare the plot for various pH values. When the GH versus I_R curve in differentiated, the K versus I relation is obtained, as mentioned in Sec. 8.2. The procedure shown in case 2 is then followed. If this method for determining K(I) is not employed, the chromatographer should obey the following procedure.

From the linear gradient elution experiments, we also determine the R_s values according to Eq. (2.51) and plot them against the following term on the basis of Eq. (8.12), $Y = (D_m I_a Z/$ $GHud_p^2)^{1/2}$. In the calculation of Y, I_a and D_m are taken to be 1.0 M and 6.7×10^{-7} cm^2/s. However, the choice of D_m is arbitrary. Thus, for example, 1×10^{-6} cm^2/s may be employed as a representative value. Although the peak position is not affected by the flow rate, R_s is varied with it. It is thus desirable that the R_s values measured at different flow rates be included in the above plots. We are now able to calculate both the peak position and the resolution from the above two charts, that is, GH versus I_R and R_s versus $Y = [D_m I_a Z/$ $(GHud_p^2)]^{1/2}$. Therefore, we give examples of how to calculate the operating and column variables for a column with different dimensions by using the two charts.

Let assume that a column 30 cm in diameter and 20 cm high is employed for the scale-up of productions. When ε is assumed to be 0.4, V_0, V_t, and A_c are 5.65 L, 14.1 L, and 707 cm^2, respectively. As mentioned in Chap. 9, it is not easy to obtain the same mobile-phase velocity u for a large column as for a small column. Let us assume that the u value appropriate for a large column is 2.0 cm/min. The flow rate F is then

$$F = uA_c \varepsilon = 2.0 \times 707 \times 0.4 = 566 \text{ mL/min} \tag{10.16}$$

When the R_s value = 1.8 is desired, the corresponding value of Y is read as 27 from Fig. 10.3. The GH and g values are then

$$GH = D_m I_a Z(Y^2 ud_p^2)$$

$$= \frac{6.7 \times 10^{-7} \times 1 \times 20}{27^2 \times (2.0/60) \times (40 \times 10^{-4})^2}$$

$$= 0.0345 \text{ M} \tag{10.17}$$

$$g = \frac{GH}{V_t - V_0} = \frac{0.0345}{14,100 - 5650} = 4.08 \times 10^{-6} \text{ M/mL} \tag{10.17'}$$

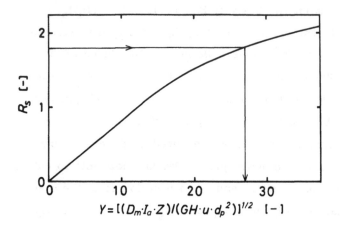

Fig. 10.3 Schematic representaion of the relation between R_s and $Y = (D_m I_a Z / GHud_p^2)^{1/2}$.

We then read out I_R corresponding to this GH from the charts (see Fig. 10.2). The obtained $I_R = 0.182$ yields the peak position for substance 1 as follows:

$$V_e = Ft_R = \frac{I_R - I_0}{g} + V'$$

$$= \frac{0.182 - 0.03}{4.08 \times 10^{-6}} + [5650 + (14100 - 5650) \times 0.9]$$

$$= 50.5 \text{ L} \tag{10.18}$$

$$t_R = \frac{V_e}{F} = \frac{50,500}{566} = 89 \text{ min} \tag{10.19}$$

In the same manner, V_e or t_R for substance 2 can be calculated. Note that t_R for substance 2 can be regarded as the separation time. When the d_p as well as the column dimension and the flow rate is changed, the GH and g is also calculated on the basis of Eqs. (10.17) and (10.17').

Case 2

When K is known as a function of I, the plots of GH versus
I can be prepared by the numerical integration of $1/(K - K')$
with respect to I as described previously. We take the same
conditions employed in case 1. That is, d_c = 30 cm, Z = 20
cm, d_p = 40 μm, ε = 0.4, GH = 0.0345 M, and u = 2.0 cm/min.
The value of I_R read from the chart is also 0.182 M. In order
to use a chart shown in Fig. 8.13 for predicting the peak width,
we must estimate the value of N_p according to Eq. (2.75):

$$N_p = \frac{Z}{2(D_L/u) + d_p^{\,2}uHK_R^{\,2}/[30K_{crt}\overline{D}_{crt}(1 + HK_R)^2]} \qquad (10.20)$$

As for the parameters in this equation, $\overline{D}_{crt}$ (= $\overline{D}$) is already estimated as 1.0×10^{-7} cm^2/s. As a good approximation,
D_L/u is assigned to be equal to d_p as discussed in Chap. 5.
The distribution coefficient K_{crt} under conditions in which the
protein is not adsorbed to the ion exchanger usually ranges
between 0.2 and 0.8, as shown in Table 3.1. Therefore, K_{crt}
is taken to be 0.5. K_R can be calculated from I_R. We assume
that the calculated K_R is 1.5. The insertion of these values
into Eq. (10.20) then yields

$$N_p = \frac{20}{2 \times 0.004 + \dfrac{0.004^2 \times (2.0/60) \times 1.5 \times 1.5^2}{[30 \times 0.5 \times 1.0 \times 10^{-7} \times (1 + 1.5 \times 1.5)^2]}}$$

$$= 164 \qquad (10.21)$$

Note that the effect of the particle diameter is also easily expected by means of this calculation.

Next we must calculate the value of the term $-(1 + HK_R)/$
$[2GH(1 + HK')(dK/dI)]$. Since K is known as a function of I,

dK/dI at $I = I_R$ can be calculated. Let us assume the calculated $dK/dI = -40$. The values of the other parameters are already known. The value of this term then becomes

$$-\frac{1 + 1.5 \times 1.5}{2 \times 0.0345 \times (1 + 1.5 \times 0.9) \times (-40)} = 0.501 \qquad (10.22)$$

The value of the vertical axis of the correlation shown in Fig. 8.13 corresponding to this value is read as 0.43. σ_θ is then calculated as

$$
\begin{aligned}
\sigma_\theta &= \left(0.43 \times \frac{(1 + HK_R)^2}{N_p} \right)^{1/2} \\
&= \left(0.43 \times \frac{(1 + 1.5 \times 1.5)^2}{164} \right)^{1/2} \\
&= 0.166
\end{aligned}
\qquad (10.23)
$$

The width of the elution curve having a unit of time is given as $4\sigma_\theta Z/u = 4 \times 0.166 \times 20/2.0 = 6.6$ min. Insertion of I_R and σ_θ thus calculated for substances 1 and 2 into Eq. (2.88) yields R_s.

The sample volume can be increased without a significant loss in resolution until it may occupy 10% of the column volume, as described in Sec. 8.1. The maximum sample volume can be easily calculated in case 1 as follows. When the distribution coefficient at the initial condition K_0 is 100, the maximum sample volume applicable to the column 30 cm in diameter and 20 cm high is given by

$$
\begin{aligned}
\text{Sample volume} &= [V_0 + K_0(V_t - V_0)]0.1 \\
&= [5650 + 100 \times (14,100 - 5650)] \times 0.1 \\
&= 85,000 \text{ mL}
\end{aligned}
\qquad (10.24)
$$

This equation means that the maximum sample volume increases with the column volume when the initial condition is the same.

On the other hand, in case 1, in which K_0 is unknown, it is advisable that several preliminary experiments with different sample volumes be carried out in order to determine the relation between the R_s and the sample volume, as shown in Fig. 8.31.

In both case 1 and case 2, when only the column diameter is increased, both the increase in the volumetric flow rate F and the decrease in the slope of the gradient g according to Eq. (10.25) give the same R_s and t_R value, provided that the large column is mechanically stable to the increase of F.

$$\frac{F_L}{F_S} = \frac{g_S}{g_L} = \frac{A_{c,L}}{A_{c,S}} \qquad (10.25)$$

The subscripts L and S in Eq. (10.25) indicate the large column and small column, respectively.

10.2 STEPWISE ELUTION

As stressed previously, the choice of the elution schedule for stepwise elution is not easy unless the distribution coefficient K of each substance contained in a sample is available as a function of ionic strength I and pH. We therefore show examples of the calculation procedure for the two cases treated in Sec. 10.2. Namely, the distribution coefficient as a function of ionic strength K(I) is unknown in case 1 but it is available in case 2. We first treat case 2.

Case 2

In stepwise elution, there are two types of elution behavior, as discussed in Sec. 8.4. When the ionic strength of an elution buffer I_{elu} is high so that the distribution coefficient of a protein in the elution buffer K_{elu} is lower than that of a salt K', a very sharp protein peak appears in the spreading boundary of the elution buffer (type I elution). On the other hand, both the peak retention time and the peak width increase with the decrease in I_{elu} when K_{elu} is higher than K' (type II elution). Obviously, the former is desirable. The ionic strength at which K of a given protein is equal to K' is referred to as I'. In order to separate a desired protein from the other proteins successfully by type I elution, while the desired protein is eluted the others must be retained near the top of the column. This means that K_{elu} of the other protein must be much higher than 1.0 at I' of the desired protein. If this condition is found, the operational variables for type I elution are easily calculated. For example, the same conditions as those used in Sec. 10.1 are chosen: d_c = 30 cm, Z = 20 cm, ε = 0.4, and K' = 0.9. The flow rate can be higher than that employed for linear gradient elution, described in Sec. 10.1, since a very sharp peak can be obtained in type I elution owing to the zone sharpening effect discussed in Sec. 8.4. The elution curve will appear after nearly one column volume of each elution buffer is introduced.

$$V_e = Ft_R = [V_0 + K'(V_t - V_0)]$$
$$= 13,300 \text{ mL} \tag{10.26}$$

However, we must start collecting effluent before V_e in the actual operation. The maximum sample volume is the same as that calculated in Sec. 10.1 when the same initial condition is employed. That is, 85,000 mL can be applied to this column [see Eq. (10.24)]. However, the sample volume can be further increased when the yield and concentration rather than purification are required.

Tyep I elution is also an efficient method for the concentration of proteins, as shown in Sec. 8.5 (see Table 8.2). In

this case, the column design should be chosen such that the
column is saturated with the sample before elution. Therefore,
when this column is employed, the sample volume is 850,000 mL.
If the sample to be concentrated is 100,000 mL, the column
length can be reduced to

$$\text{Column length} = \frac{100,000}{A_c \varepsilon (1 + HK_0)}$$

$$= \frac{100,000}{707 \times 0.4 \times (1 + 1.5 \times 100)}$$

$$= 2.3 \text{ cm} \qquad\qquad (10.27)$$

Actually, the front boundary of the sample zone spreads owing
to gel-phase diffusion and axial dispersion. Therefore, the
column length must be higher than this value in order to retain
all the sample within the column. However, it is noteworthy
that a few centimeters of column is sufficient for the concentration of proteins by IEC. After the sample is applied to the
column, the elution is performed with a high-ionic-strength
buffer. The concentration of the desorbed fraction becomes
higher with decreasing flow rate, which increases the number
of plates. Since the column is very short, the elution does not
take much time even at low flow rates. The fraction is collected
until the elution volume becomes equal to two to three column
volumes. The sample is usually concentrated by a factor of
10 or more (see Table 8.2).

However, type I elution is applicable only when the properties of two proteins are sufficiently different. If a suitable
condition is not found, type II elution must be employed. In
this case, the peak position can be calculated as

$$t_R = \frac{Z}{u} (1 + HK_{elu}) \qquad\qquad (10.28)$$

For estimation of the peak width, N_p should be calculated in the same manner as described in Sec. 10.1 [see Eqs. (10.20) and (10.21)]. In this calculation, K_R must be read as K_{elu}. The peak width having a dimension of time W is then given as $4(Z/u)(1 + HK_{elu})/N_p^{1/2}$. The R_s value can be calculated by insertion of the t_R and W values for two proteins into Eq. (2.51). Usually we search the ionic strength at which the difference of K_{elu} of two proteins $K_{elu,2} - K_{elu,1}$ is large to obtain a high R_s. However, since the increase in K_{elu} causes an increase in t_R, an increase in the peak width, and a decrease in the peak height, it is not practical that K_{elu} be larger than 5.

Case 1

When K is unknown, we again employ the plots of GH and I_R prepared from linear gradient elution experiments, as described in the previous section. A schematic representation of the GH versus I_R plots is again shown in Fig. 10.4. When K(I) is obtained from differentiation of the GH versus I_R curve, as shown in Sec. 8.2, the same procedure as that in case 2 can be followed. A more simplified method for the rough estimation of the chromatographic conditions in stepwise elution is presented here.

As stated in Sec. 8.2, I_R increases no more when GH is increased up to a certain value. This value of I_R is equal to I', as mentioned in Sec. 8.4. Therefore, if the value of GH for the other proteins is much lower at I' of a given protein, it will be eluted with the elution buffer of I_{elu} = I' in a very concentrated fraction while the other proteins are retained in the column (type I elution). In the same manner, the second elution buffer is chosen from the chart. This procedure is depicted in Fig. 10.4.

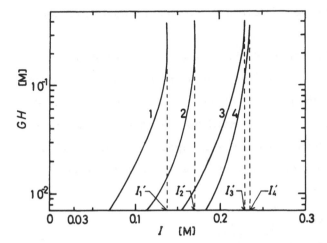

Fig. 10.4 Schematic representation of the relation GH versus I_R. The numbers 1, 2, 3, and 4 imply substances 1, 2, 3, and 4, respectively.

When the curves of GH versus I_R are not sufficiently different (this is usually the case in which the number of components
are contained in the sample), the foregoing simple procedure
is not applicable. The separation of substances 3 and 4 in Fig.
10.4 is a typical example of this case. As stated in Sec. 8.2,
the difference in the GH values at certain I_R corresponds implicitly to that of the K values. Therefore, the ionic strength
of the elution buffer I_{elu} to achieve a good separation can be
estimated from the GH versus I_R curve so that the GH values of
the two proteins at $I_R = I_{elu}$ are sufficiently different. However, we also should keep in mind that if the value of GH at
this I_R is too low, for example, less than 0.01, the resulting
elution curve is diluted and widened significantly. For estimation of the peak position, it is recommended that preliminary
experiments be carried out with a small column. In addition,

information on the peak width can also be obtained from such experiments.

When the elution schedule is determined by either of the preceding two procedures, the change in the column and operating variables without a loss of column efficiency obtained with a given set of experimental conditions can be easily calculated. When only the column diameter is changed, the sample volume and the volumetric flow rate should be changed in proportion to the column cross-sectional area:

$$\frac{F_L}{F_S} = \frac{(\text{sample volume})_L}{(\text{sample volume})_S} = \frac{A_{c,L}}{A_{c,S}} \qquad (10.29)$$

In addition, if the column length is varied, we must adjust the separation efficiency by controlling the flow rate. For example, when the column length is shortened to two-thirds that of the small column, the resolution decreases by a factor of the square root of two-thirds = 0.816 in type II elution. To compensate for this decrease in resolution, the flow rate must be decreased according to the calculation procedure described previously [see Eqs. (10.20) and (10.21)] so that N_p becomes the same as that of the smaller but longer column.

The calculation procedures described in this chapter do not take into account the effect of initial concentration on the elution behavior, as described in Sec. 8.1. Therefore, if the concentration of proteins contained in a sample is high, we must incorporate the loss of resolution and the shift in peak position into the calculation. Unfortunately, such a calculation is not easy, as already stated. However, even in such cases we recommend that linear gradient elution experiments be carried out with a diluted fraction or a small amount of the sample as

preliminary experiments. The results obtained may be interpreted as mentioned in this chapter and will serve as fundamental data. It is further assumed here that the column efficiency
does not vary when the column dimension is changed.

CONCLUDING REMARKS AND FUTURE ASPECTS

Although it was in the early 1960s that *ion-exchange chromatorgaphy*
(IEC) was recognized as a general method for protein separation,
it has been developed drastically especially in the last decade.
It is no exaggeration to say that the separation steps of almost
all proteins include IEC. In addition to the conventional IEC
packing materials, which have a low mechanical stability, various
rigid and small IEC packings have been recently developed and
are commercially available. These IEC packings not only increase
the speed of separation and the resolution but also make large
scale IEC easy. Process scale *high-performance liquid chromatography* (HPLC) is already carried out in the case of the reversed-phase separation mode, which can process kilograms of
crude samples within an hour (Dwyer, 1984). It is expected that
process scale HPIEC packed with particles of 20 to 40 µm in
diameter is employed in the downstream processes in bioindustries
(Regnier and Gooding, 1980; Janson and Hedman, 1982; Unger
and Janzen, 1986).

On the other hand, analytical HPLC packings and columns
will become much smaller. The advantages in the use of micro
HPLC columns have been discussed previously (Scott, 1980;
Novotony, 1981; Yang, 1983). As mentioned in the text, because
of its high resolution, high speed of separation, and high sensitivity, such micro HPIEC can be used in testing the homogeneity of proteins as an alternative of the conventional disc gel
electrophoresis method, in on-line monitoring of the bioproduction
process, or in analyzing a very small amount of proteins (sub-
micrograms).

As for the theoretical aspects of IEC of proteins, the separation mechanism is clarified qualitatively. However, as already
pointed out in this book, preliminary estimation of values of
many important parameters appearing in the theoretical model is
not easy. We presented several methods for estimating such
parameters as gel phase diffusivity and axial dispersion coefficient. Although the ionic strength or pH dependent distribution coefficient between proteins and ion exchangers are required

for the prediction of both the retention and the resolution, its theoretical prediction is difficult since the exact adsorption mechanism, that is, ion exchange equilibrium, is still unknown as already mentioned. A simple method for predicting the lowering of the column efficiency with increasing sample loading is also needed, since in the preparative IEC the sample overloading may be allowable in order to increase the throughput.

In general, the downstream process of proteins and enzymes includes several chromatographic methods such as *gel filtration chromatography* (GFC) and biospecific *affinity chromatography* (AFC) as well as IEC. These various types of chromatography are performed sequentially. However, it is still difficult to determine the suitable sequence of different chromatographic methods for the purification of a given protein. So, we must establish methods for designing chromatographic separation processes including liquid chromatography of various different separation modes. This can increase not only the yield of the end-product but also the degree of purification. The time needed for the separation is also reduced by such optimization methods.

It seems that IEC may be replaced by AFC, which has a high specific affinity to a given protein. For example, AFC with immobilized monoclonal antibodies is probably the most efficient LC at present. However, since the affinity is too strong, the desorption must be performed with extremely low or high pH solutions and/or with protein denaturants such as urea and guanidine hydrochloride. Moreover, the preparation of monoclonal antibodies immobilized packings is laborious and time-consuming, which results in extremely high costs of the packings (Chase, 1984). On the other hand, in the case of IEC, any protein can be adsorbed to either of cation or anion exchange columns at a certain pH and salt concentration. The elution can be performed at mild conditions, where a desired resolution is achieved and the yield (recovery) is usually high. In addition, various types of superior IEC packings and columns have continuously been developed, which are not so expensive as AFC packings. For these reasons, IEC will be playing an important role in the separation and purificaiton of proteins, enzymes and other related materials.

The improvement and/or development of novel chromatographic systems such as those shown in Sec. 8.5 are also an important subject. If a much better system than the conventional system is developed, it may cause a revolution in the field of chromatography. New efficient separation methods like those discussed in Sec. 8.5 may also change the state of the art rapidly.

Appendix

A. DERIVATION OF THE VALUE OF K_s

We derive the value of K_s defined by Eq. (A.1) from the diffusion equation (A.2) (for the sake of simplicity the overbar employed in the text is dropped).

$$\frac{\partial C_{av}}{\partial t} = K_s (C_s - C_{av}) \tag{A.1}$$

$$\frac{\partial C}{\partial t} = D \left(\frac{\partial^2 C}{\partial r^2} + \frac{2}{r} \frac{\partial C}{\partial r} \right) \tag{A.2}$$

Integration of Eq. (A.2) with respect to r from $r = 0$ to $r = R_p$ yields

$$\frac{\partial C_{av}}{\partial t} = E \left(\frac{\partial C}{\partial r} \right)_{r = Rp} \tag{A.3}$$

where $C_{av} = (3/R_p^3) \int_0^{Rp} r^2 C \, dr$ and $E = 3D/R_p$. The solution of Eq. (A.2) in the Laplace domain is

$$\tilde{C} = \tilde{C}_s \, \frac{R_p}{r} \, \frac{\sinh (f)}{\sinh (F)} \tag{A.4}$$

where $f = r(p/D)^{1/2}$

$F = R_p(p/D)^{1/2}$

p = Laplace transform variable

Then,

$$\left(\frac{d\tilde{C}}{dr}\right)_{r=R_p} = \frac{\tilde{C}_s}{R_p} \, G \tag{A.5}$$

and

$$\tilde{C}_{av} = \frac{3\tilde{C}_s G}{F^2} \tag{A.6}$$

where

$$G = F \coth (F) - 1 \tag{A.7}$$

From Eqs. (A.1) and (A.3), K_s at infinite time is given by

$$K_s = \lim_{t \to \infty} \frac{E \, (\partial C/\partial r)_{r=R_p}}{C_s - C_{av}} \tag{A.8}$$

We employ a property of the Laplace transform that $\lim_{t \to \infty} f(t) = \lim_{p \to 0} p\tilde{f}(p)$. Equation (A.8) then becomes

$$K_s = \lim_{p \to 0} \frac{pE \, (d\tilde{C}/dr)_{r=R_p}}{p(\tilde{C}_s - \tilde{C}_{av})} = \frac{E}{R_p} \lim_{p \to 0} \frac{G}{1 - 3G/F^2} \tag{A.9}$$

By expanding coth (F) in Eq. (A.7), G is expanded as

$$G = \frac{F^2}{3} - \frac{F^4}{45} + \cdots \tag{A.10}$$

$$K_s = \lim_{p \to 0} \frac{E}{R_p} \frac{F^2/3 - F^4/45}{1 - 1 + F^2/15}$$

$$= \lim_{p \to 0} \frac{3D}{R_p^2}\left(5 - \frac{F^2}{3}\right) = \frac{15D}{R_p^2} = \frac{60D}{d_p^2}$$

The value of K_s is then

$$K_s = \frac{15D}{R_p^2} = \frac{60D}{d_p^2} \tag{A.11}$$

B.   DERIVATION OF EQ. (2.59)

Equation (2.58) is rearranged in the reduced form as

$$\frac{dC_{(n)}}{d\theta} + \frac{Hd\overline{C}_{(n)}}{d\theta} = N_p[C_{(n-1)} - C_{(n)}] \tag{B.1}$$

(see p. 363 for the meaning of the symbols).

Since $\overline{C}_{(n)}$ is a function of both C and I,

$$\frac{d\overline{C}_{(n)}}{d\theta} = \frac{d\overline{C}_{(n)}}{dI}\frac{dI}{d\theta} + \frac{d\overline{C}_{(n)}}{dC_{(n)}}\frac{dC_{(n)}}{d\theta} \tag{B.2}$$

Assumption 3 in Sec. 2.2.2 permits the insertion of Eq. (2.56) into Eq. (B.2). The first term on the right-hand side of Eq. (B.2) then becomes

$$\frac{d\overline{C}_{(n)}}{dI}\frac{dI}{d\theta} = \frac{d\{K[C_{(n)}, I]C_{(n)}\}}{dI}\frac{dI}{d\theta} = \frac{dI}{d\theta}\frac{dK[C_{(n)}, I]}{dI}C_{(n)} \tag{B.3}$$

The second term is represented by

$$\frac{d\overline{C}_{(n)}}{dC_{(n)}}\frac{dC_{(n)}}{d\theta} = \frac{dC_{(n)}}{d\theta}\frac{d\{K[C_{(n)}, I]C_{(n)}\}}{dC_{(n)}}$$

$$= K[C_{(n)}, \ I] \ \frac{dC_{(n)}}{d\theta} + \frac{dC_{(n)}}{d\theta} \ \frac{dK[C_{(n)}, \ I]}{dC_{(n)}} \ C_{(n)}$$

$$(B.4)$$

Substitution of Eqs. (B.2) through (B.4) into Eq. (B.1) gives Eq. (2.59).

C.   ANALYTICAL SOLUTIONS FOR A CHANGE IN IONIC STRENGTH IN STEPWISE ELUTION AND LINEAR GRADIENT ELUTION

C.1   Stepwise Elution

The basic equation (2.60) is rewritten using the dimensionless parameter $I^*_{(n')} = [I_{(n')} - I_0]/[I_{elu} - I_0]$:

$$\frac{dI^*_{(n')}}{d\theta} + R'I^*_{(n')} = R'I^*_{(n'-1)} \tag{C.1}$$

where $I^*_{(0)} = 1$ and the initial condition is $I^*_{(1)} = I^*_{(2)} = \cdots = I^*_{(N_p')} = 0$. We then use the Laplace transform method to obtain the solution.

Equation (C.1) is transformed and becomes, for $n' = 1$,

$$p\tilde{I}^*_{(1)} + R'\tilde{I}^*_{(1)} = R'\tilde{I}^*_{(0)} = \frac{R'}{p} \tag{C.2}$$

where $\tilde{I}^* =$ transformed I^* and $p =$ Laplace transform variable. Equation (C.2) is solved for $\tilde{I}^*_{(1)}$ as

$$\tilde{I}^*_{(1)} = \frac{R'}{p(p + R')} \tag{C.3}$$

Repetition of this operation yields

$$\tilde{I}^*_{(n')} = \frac{R'^{n'}}{P(P + R')^{n'}} = \tilde{I}^*_{(n'-1)} - \frac{R'^{n'-1}}{(P + R')^{n'}} \tag{C.4}$$

The following relation is available from standard tables for the Laplace transform:

$$\mathcal{L}^{-1} \frac{1}{(p + R')^{n'}} = \frac{\theta^{n'-1}}{(n' - 1)!} e^{(-\theta R')} \tag{C.5}$$

where $\mathcal{L}^{-1}$ = inverse Laplace transform operation. The inverse transformation of Eq. (C.4) with the aid of Eq. (C.5) gives

$$I^*_{(n')} = I^*_{(n'-1)} - g'(\theta, n') = f(\theta, n') \tag{C.6}$$

$$f(\theta, n') = 1 - g(\theta, n') \tag{C.7}$$

$$g(\theta, n') = \sum_1^{n'} g'(\theta, n') \tag{C.8}$$

$$g'(\theta, n') = \frac{(R'\theta)^{n'-1}}{(n' - 1)!} e^{-\theta R'} \tag{C.9}$$

The expression for $dI^*_{(n')}/d\theta$ is given by a property of the Laplace transform:

$$\frac{dI^*_{(n')}}{d\theta} = \mathcal{L}^{-1}[pI^*_{(n')}] = \mathcal{L}^{-1}\left[\frac{R'^{n'}}{(P + R')^{n'}} \right] = R'g'(\theta, n') \tag{C.10}$$

C.2   Linear Gradient Elution

Introduction to the parameter $I^*_{(n')} = [I_{(n')} - I_0]R'/G$ gives the same basic equation as Eq. (C.1) and the intial condition is also the same as in stepwise elution. Since $I^*_{(0)} = R'\theta$, $\tilde{I}^*_{(0)} = R'/p^2$.

Similar operations in stepwise elution yield the solution for n' in the Laplace domain as

$$\tilde{I}^*_{(n')} = \tilde{I}^*_{(n'-1)} - \frac{R'^{n'}}{P(P + R')^{n'}} = \frac{R'^{n'+1}}{P^2(P + R')^{n'}} \tag{C.11}$$

The second term on the right-hand side is the same as for Eq.
(C.4). Therefore, $I^*_{(n')}$ is given by

$$I^*_{(n')} = I^*_{(n'-1)} - f(\theta, n') = R\theta - \sum_1^{n'} f(\theta, n') \qquad (C.12)$$

$dI^*_{(n')}/d\theta$ is given by

$$\frac{dI^*_{(n')}}{d\theta} = \mathcal{L}^{-1}[p\tilde{I}^*_{(n')}] = \mathcal{L}^{-1}\left[\frac{R'^{n'+1}}{P(P + R')^{n'}}\right] = R'f(\theta, n')$$

$$(C.13)$$

Conversion of I* to I in Eqs. (C.6), (C.10), (C.12), and
(C.13) gives the solution in Sec. 2.2.2.

D. NUMERICAL CALCULATION PROCEDURE

Equation (2.59) is numerically calculated by the Runge-Kutta-
Gill method (Lapidus, 1962), which has been the most widely
used formula for numerical calculation of ordinary differential
equations. Equation (2.59) is rewritten by using the function
f_n as $f_n = dC_{(n)}/d\theta = f_n[I, dI/d\theta, K, dK/dI, C_{(n)}, C_{(n-1)}]$.
In the Runge-Kutta-Gill method, the function f_n must be evalua-
ted four times per time step. N'_p is usually greater than N_p, so
for I and $dI/d\theta$ in f_n, the values of $I_{(n')}$ and $dI_{(n')}/d\theta$ where n'
corresponds to the position of the center of the nth plate, are
employed. Here, $I_{(n')}$ and $dI_{(n')}/d\theta$ are calculated by the
analytical solutions in the text at each time level. For a large
value of $(n' - 1)!$ in Eq. (2.64), Stirling's formula is employed.
Also, K and dK/dI in the function f_n are calculated at each
time level, with I corresponding to that time level and the prior
value of $C_{(n)}$. The mass of solute applied to the column C_0X_0

is checked by integrating the concentration at the outlet of the column with respect to θ, that is, $\int_0^\infty C_{(N_p)} \, d\theta$. If this integrated mass is appreciably different from $C_0 X_0$, the smaller time increment is employed. In general, increasing N_p requires a smaller time increment.

E. A SOLUTION FOR THE PEAK WIDTH IN LINEAR GRADIENT ELUTION UNDER THE QUASI-STEADY STATE

We first introduce the following new coordinates:

$$z_r = z - \frac{ut}{1 + HK_R} = z - \frac{Z\theta}{1 + HK_R} \tag{E.1}$$

$$\theta_r = \theta \tag{E.2}$$

where z_r = distance from the peak position of the protein zone, which is negative behind the peak position and positive ahead of it. The position of the nth plate is given by

$$z = \frac{nZ}{N_p} \tag{E.3}$$

Then,

$$z_r = \frac{n_r Z}{N_p} \tag{E.4}$$

The number of plates n_r is numbered from the plate at the peak position. From Eqs. (E.1), (E.3), and (E.4), the following coordinate is obtained:

$$n_r = n - S\theta \tag{E.5}$$

where

$$S = \frac{N_p}{1 + HK_R} \tag{E.6}$$

By using these coordinates, $dC_{(n)}/d\theta$ is given by

$$\frac{dC_{(n)}}{d\theta} = \frac{dC_{(n_r)}}{d\theta_r}\frac{d\theta_r}{d\theta} + \frac{dC_{(n_r)}}{dn_r}\frac{dn_r}{d\theta} = \frac{dC_{(n_r)}}{d\theta_r} - S\frac{dC_{(n_r)}}{dn_r}$$

(E.7)

Similarly,

$$\frac{d\overline{C}_{(n)}}{d\theta} = \frac{d\overline{C}_{(n_r)}}{d\theta_r} - S\frac{d\overline{C}_{(n_r)}}{dn_r}$$

(E.8)

Substitution of $\overline{C}_{(n)}$ by $KC_{(n)}$ gives

$$\frac{d\overline{C}_{(n)}}{d\theta} = K\frac{dC_{(n)}}{d\theta} + C_{(n)}\frac{dK}{d\theta}$$

(E.9)

$$\frac{d\overline{C}_{(n)}}{dn} = K\frac{dC_{(n)}}{dn} + C_{(n)}\frac{dK}{dn}$$

(E.10)

(Hereafter, the subscripts r for n_r and θ_r are dropped for convenience.)

Since a steady state with respect to θ is assumed, $dC_{(n)}/d\theta$ and $dK/d\theta$ are set equal to zero. Insertion of Eqs. (E.7) through (E.10) into Eq. (B.1) in Appendix B gives

$$S(1 + HK)\frac{dC_{(n)}}{dn} + HSC_{(n)}\frac{dK}{dn} - N_p[C_{(n)} - C_{(n-1)}] = 0$$

(E.11)

$dC_{(n)}/dn$ in Eq. (E.11) is then replaced by the central difference approximation; that is, $dC_{(n)}/dn = [(C_{(n+1)} - C_{(n-1)}]/2$

$$S(1 + HK)[C_{(n+1)} - C_{(n-1)}] + 2HSC_{(n)}\frac{dK}{dn}$$
$$- 2N_p[C_{(n)} - C_{(n-1)}] = 0$$

(E.12)

This equation is rearranged as

$$H(K - K_R)[C_{(n+1)} - C_{(n-1)}] + (1 + HK_R)[(C_{(n+1)} + C_{(n-1)}$$

$$- 2C_{(n)}] + 2HC_{(n)} \frac{dK}{dn} = 0 \qquad (E.12)$$

We introduce the following approximate relation for $K - K_R$:

$$K - K_R = \frac{dK}{dI} \left(\frac{dI}{dz_r}\right) z_r = \frac{dK}{dI} \left(\frac{dI}{dn}\right) n$$

Equation (E.12) is then rewritten as

$$an[C_{(n+1)} - C_{(n-1)}] + b[C_{(n+1)} + C_{(n-1)} - 2C_{(n)}]$$

$$+ 2aC_{(n)} = 0 \qquad (E.13)$$

where $a = H(dK/dI)(dI/dn)$ and $b = 1 + HK_R$. This equation is solved for $C_{(n+1)}$ as

$$C_{(n+1)} = \frac{2(1 - A)C_{(n)} + (An - 1)C_{(n-1)}}{1 + An} \qquad (E.14)$$

where $A = a/b$. For $n = 0$,

$$C_{(1)} = 2(1 - A)C_{(0)} - C_{(-1)} \qquad (E.15)$$

The concentration distribution is considered symmetrical with its maximum at $n = 0$. Therefore, $C_{(1)} = C_{(-1)}$ and $C_{(0)}$ has an arbitrary maximum concentration, say, C_{max}. Then,

$$C_{(1)} = C_{max}(1 - A) \qquad (E.16)$$

$$C_{(2)} = \frac{C_{max}(1 - A)(1 - 2A)}{1 + A} \qquad \text{for } n = 1 \qquad (E.17)$$

$$C_{(3)} = \frac{C_{max}(1 - A)(1 - 2A)(1 - 3A)}{(1 + A)(1 + 2A)} \qquad \text{for } n = 2 \qquad (E.18)$$

$$C_{(4)} = \frac{C_{max}(1 - A)(1 - 2A)(1 - 3A)(1 - 4A)}{(1 + A)(1 + 2A)(1 + 3A)} \qquad \text{for } n = 3$$

$$\text{(E.19)}$$

Repetition of this operation yields

$$C_{(n)} = \frac{C_{max}(1 - A)(1 - 2A) \cdots (1 - nA)}{1(1 + A)(1 + 2A) \cdots [1 + (n - 1)A]} \qquad \text{(E.20)}$$

Since A is small, $C_{(n)}$ becomes

$$C_{(n)} = C_{max}(1 - nA)^n = C_{max}[(1 - nA)^{(1/An)}]^{(AN)}$$

$$= C_{max} \, e^{[-AN]} = C_{max} \, e^{-1/[2(1/2A)]} \qquad \text{(E.21)}$$

where $N = n^2$.

Equation (E.21) is asymptotic solution for the concentration distribution in the column, which has a form of gaussian distribution with its variance σ_n^2 given by

$$\sigma_n^2 = \frac{1}{2A} = \frac{b}{2a} = \frac{1 + HK_R}{2H \ (dK/dI) \ (dI/dn)} \qquad \text{(E.22)}$$

σ_n^2 is then converted from the dimensionless length unit to the dimensionless time unit σ_θ^2 by multiplying $(1 + HK_R)^2/N_p^2$ by Eq. (E.22):

$$\sigma_\theta^2 = \frac{(1 + HK_R)^3}{2N_p^2 \ (dK/dI) \ (dI/dn) \ H} \qquad \text{(E.23)}$$

here, dI/dn in Eq. (E.23) is considered constant and is obtained by differentiation of Eq. (2.70) in Sec. 2.2.2 with respect to n:

$$\frac{dI}{dn} = -\frac{G(1 + HK')}{N_p} \qquad\qquad (E.24)$$

After Eq. (E.24) is inserted into Eq. (E.23), Eq. (E.23) is rearranged as Eq. (2.86) in Sec. 2.2.3.

F. VALUES OF THE MOLECULAR DIFFUSION COEFFICIENT AND ITS ESTIMATION METHOD

As discussed in the text, the molecular diffusion coefficient D_m is an important parameter in the zone spreading mechanism in liquid chromatography. Values of D_m for several standard proteins, together with those for typical small molecules, such as amino acids and sugars, are listed in Table F.1. As for the estimation of D_m for proteins, Young et al. (1980) have presented the equation

$$\frac{D_m \eta}{T} = 8.34 \times 10^{-8} \, MW^{-1/3} \qquad\qquad (F.1)$$

where η = viscosity of the solvent

   T = temperature

   MW = molecular weight of proteins

Figure F.1 shows relation between $D_m \eta / T$ and MW for several proteins whose D_m values are available in the literature. It is seen from the figure that most of the data, except that for fibrinogen and myosin, are well correlated by Eq. (F.1). The reason for the deviation of the data for fibrinogen and myosin is that they are not globular proteins. It may be concluded that the correlation equation (F.1) or the correlation shown in Fig. F.1 is very useful for the estimation of D_m, especially for engineering purposes.

Table F.1 Values of D_m and MW for Several Salts, Amino Acids, Sugars, and Proteins[a]

Values at 25°C	MW	$D_m \times 10^6$ (cm²/s)	Reference[c]	Values at 20°C	MW	$D_m \times 10^6$ (cm²/s)	Reference[c]
Amino acids				Proteins			
Glycine	75	10.6	1	Ribonuclease	13,000	1.3	6
Alanine	89	9.3	1	Myoglobin	17,000	1.1	6
				α-Chymotrypsinogen	24,000	0.95	6
Sugars				β-Lactoglobulin	37,000	0.78	7
Glucose	180	6.8	2	Ovalbumin	45,000	0.74	7
Sucrose	342	5.3	3	Bovine serum albumin	68,000	0.61	7
Maltose	342	5.3	4	α-Amylase	97,000	0.57	6
Raffinose	504	4.3	1	γ-Globulin	150,000	0.48	6
				Catalase	250,000	0.41	6
Salts				Urease	480,000	0.35	6
NaCl	58	16	5	Thyroglobulin	670,000	0.26	7
(NH₄)₂SO₄	132	15	5				

[a]These D_m values can be corrected for other temperatures according to the relation that $D_m \eta/T$ is constant if the temperature is not too far from 20°C (or 25°C) and the molecule conformation is not varied with temperature.

[b]The D_m values of amino acids, sugars, and salts vary with concentration. Therefore, the following D_m values are employed for the calculation in Fig. 4.2: glycine, 7.9×10^{-5}; arginine, 5.2×10^{-6}; valine, 7.1×10^{-6}; glucose, 5.0×10^{-6}; maltose, 4.5×10^{-6}; triose, 3.8×10^{-6}; NaCl, 13.5×10^{-6} (Namikawa et al., 1977; Nakanishi et al., 1977a).

[c]References are as follows: (1) Hodgman (1966), (2) Gladden and Dole (1953), (3) Henrion (1964), (4) International Critical Tables (1929), (5) Robinson and Stokes (1955), (6) Smith (1968), (7) Kuntz and Kauzmann (1974).

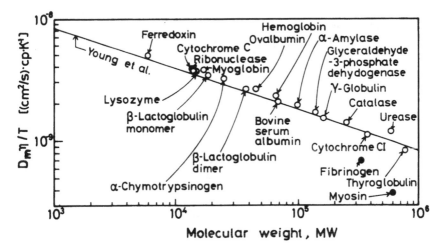

Fig. F.1 Correlation between $D_m \eta/T$ and molecular weight MW. D_m = molecular diffusion coefficient (cm^2/s), η = viscosity of solvent (cp), and T = absolute temperature (K). The straight line is calculated based on Eq. (F.1). ($\circ$, $\bullet$) Calculated from the D_m values compiled in Smith (1968) and Kuntz and Kauzmann (1974).

Symbols

a	H (dK/dI) (dI/dn) in App. E
a	parameter in Eq. (3.57) (L atm^{-1} mol^{-1})
a_i	activity of species i (M)
A	term in HETP equation (2.43) (cm)
A	parameter in Eq. (3.39)
A	a/b in Appendix E
A_c	cross-sectional area of the column (cm^2)
A_1	cross-sectional area of the mixing vessel shown in Fig. 6.4 (cm^2)
A_2	cross-sectional area of the reservoir shown in Fig. 6.4 (cm^2)
b	$1 + HK_R$ in Appendix E
b	parameter in Eq. (3.57) (L mol^{-1})
B	term in HETP equation (2.43) (cm^2/s)
B	parameter in Eq. (3.39)
C	concentration of protein at the exit of the column (% or M)
C	concentration of solute in the mobile phase (% or M)

$\overline{C}$ concentration of protein in the stationary phase (% or M)

C term in Eq. (2.43) (min or s)

$\overline{C}_{av}$ $3/R_p^3 \int_0^{R_p} r^2 \overline{C}\, dr$, average value of $\overline{C}$ (% or M)

C_i concentration of species i (% or M)

$\overline{C}_i$ concentration of species i in the ion exchanger (% or M)

C_{max} maximum peak height of the elution curve (% or M)

$C_p^{\ p}$ protein concentration in the pore liquid (% or M)

$C_p^{\ q}$ protein concentration in the solid phase (% or M)

C_{P0} initial protein concentration in the solution in Sec. 4.2.1 (% or M)

C_{Pe} equilibrium protein concentration in the solution in Sec. 4.2.1 (% or M)

C_s concentration at the interface between the particle and the outer solution (% or M)

$\overline{C}_s$ KC_s (see Fig. 2.6) (% or M)

C_0 initial concentration of protein applied to the column (% or M) or initial concentration of exchanging ion in solution in Eq. (3.36) (% or M)

$C_{(n)}$ concentration of protein at plate n in the mobile phase (% or M)

$\overline{C}_{(n)}$ concentration of protein at plate n in the stationary phase (% or M)

$C_{(0)}$ feed concentration (% or M)

C^* C/C_{ref} (C_{ref}, any fixed concentration)

$\overline{C}*$ $\overline{C}/C_{ref}$ (C_{ref}, any fixed concentration)

d $2r$ (cm)

d_c column diameter (cm)

d_p particle diameter (cm or μm)

d_{pa} mean particle diameter defined by $[\int_0^\infty d_p^5$
                         $f(d_p)\ d(d_p)/\int_0^\infty d_p^3\ f(d_p)\ d(d_p)]^{1/2}$ Eq.
                         (5.11) (cm or μm)

$d*$ d/d_p

D diffusion coefficient (cm^2/s)

$\overline{D}$ gel (stationary) phase diffusion coefficient
            (cm^2/s)

D_i diffusion coefficient of species i (cm^2/s)

$\overline{D}_i$ gel (stationary)-phase diffusion coefficient
            of species i (cm^2/s)

$\overline{D}_{PB}$ diffusion coefficient or protein defined by
            Eq. (4.8). (cm^2/s)

$\overline{D}_p^{\ p}$ diffusion coefficient of protein with respect
            to the pore liquid (cm^2/s)

D_m molecular diffusion coefficient (cm^2/s)

D_L effective longitudinal dispersion coefficient
            (cm^2/s)

$\overline{D}_{crt}$ gel-phase diffusion coefficient obtained by
            the moment method at high ionic strength
            (cm^2/s)

$\overline{D}_R$ gel-phase diffusion coefficient at $I = I_R$
            (cm^2/s)

$\overline{D}_{P,crt}$ D_{crt} for protein (cm^2/s)

$\overline{D}'$ gel-phase diffusion coefficient for salt (cm^2/s)

E $3D/R_p$ (cm/s)

E_{Don} Donnan potential (J/C)

f	$r(p/D)^{1/2}$
f	particle size distribution function in Eq. (5.11) (cm^{-1})
f	function $f(Z, p)$ in Eq. (3.43)
F	volumetric flow rate $(cm^3/s$ or mL/min)
F	$R_p(p/D)^{1/2}$ in App. A
F_2	flow rate of a pump shown in Fig. 6.6 (cm^3/s)
$\mathfrak{F}$	Faraday constant (C/mol)
g	$(I_f - I_0)/V_G$ slope of the gradient (M/mL)
g	function $g(Z, p)$ in Eq. (3.44)
G	$F \coth (F) - 1$ in App. A
G	gV_0 slope of the gradient normalized with respect to the column void volume (M)
GH	$g(V_t - V_0)$ slope of the gradient normalized with respect to the column gel volume (M)
h	$F_2/(F - F_2)$ or $= A_2/A_1$ in Chap. 6
h	$HETP/d_p$, reduced HETP
H	$(1 - \varepsilon)/\varepsilon = (V_t - V_0)/V_0$
HETP	height equivalent to a theoretical plate (cm)
I	ionic strength (M)
I_a	arbitrary ionic strength in Eq. (2.82) or a dimensional constant having a numerical value of 1 in the term Y (M)
I_{crt}	ionic strength above which the ion-exchange gels act as gel filtration media (M)
I_{elu}	ionic strength of the elution buffer in step-wise elution (M)
I_R	ionic strength at θ_R (M)
I_0	ionic strength with which column is initially equilibrated (M)

I_i	I of the initial starting buffer solution contained in the mixing vessel of the apparatus shown in Fig. 6.4 (M)
I_f	I of the final buffer solution in the reservoir for the apparatus shown in Fig. 6.4. (M)
$I_{(n')}$	I at plate n' in the mobile phase (M)
$I^*_{(n')}$	$[I_{(n')} - I_0]/[I_{elu} - I_0]$ or $= [I_{(n')} - I_0]/[R'/G]$ dimensionless ionic strength at plate n' in the mobile phase
I^*	$(I - I_0)/(I_{elu} - I_0)$, dimensionless ionic strength
I'	I at which $K = K'$ (M)
J_i	flux of species i $[mol/(cm^2 \cdot s)]$
$(J_i)_{diff}$	pure (statistical) diffusion flux due to the concentration gradient $[mol/(cm^2 \cdot s)]$
$(J_i)_{el}$	flux caused by electric potential gradient $[mol/(cm^2 \cdot s)]$
k	empirical constant in Eq. (3.36) (mol/g)
k'	HK, capacity factor
k_f	mass transfer coefficient (m/s or cm/s)
K	$\overline{C}/C$ or $\overline{m}_P/m_P$, distribution coefficient
K_a	adsorption rate constant (s^{-1})
K_B	$\overline{m}_B/m_B$ distribution coefficient of species B
K_{Cl}	association constant defined by Eq. (3.59) (M^{-1})
K_{crt}	K at high ionic strengths in which the ion-exchange gels act as gel filtration media
K_d	desorption rate constant (s^{-1})
K_1	equilibrium constant
K_{elu}	distribution coefficient at $I = I_{elu}$

K_f $k_f(3/R_p)/K$ mass transport coefficient defined by Eq. (2.23) (s^{-1})

K_n generalized mass transfer coefficient defined by Eq. (2.32) (n = d, f, s, or t) (s^{-1})

K_{peak} K determined from the peak position

K_p association constant defined by Eq. (3.58) (M^{-1})

K_{ps} $\overline{C}_p{}^p/C_p$

K_{pq} $\overline{C}_p{}^q/\overline{C}_p{}^q$

K_R distribution coefficient at $I = I_R$

K_s mass transfer coefficient defined by Eq. (2.30) (s^{-1})

K_t $1/[(1/K_f) + (1/K_s)]$, overall mass transfer coefficient (s^{-1})

K_0 K at $C = C_0$

K_1 parameter expressing a concentration dependence of K (see Fig. 2.4)

K' distribution coefficient for salt

L   concentration of rod (dextran molecule) (cm rod/cm^3)

m_i molarity of species i (M)

$\overline{m}_i$ molarity of species i in the ion exchanger (M)

$m_p{}^f$ molarity of the protein in the outer solution at equilibrium (M)

$m_p{}^i$ molarity of the protein in the outer solutions at the initial state (M)

M $(C_{P0} - C_P)/(C_{P0} - C_{Pe})$

M   mol per liter

MW   molecular weight

n	number of plates numbered from the top of the column
n_r	number of plates numbered from the peak position of the protein zone
n_i	number of moles of species i
$\overline{n_i}$	number of moles of species i in the ion exchanger
N_{eff}	number of effective plates defined by Eq. (2.50)
N_p	total number of theoretical plates
N'_p	N_p for salt
p	empirical constant in Eq. (3.36)
p	$\alpha + \beta i$, Laplace transform variable (s^{-1})
P, $\overline{P}$	pressure in the outer solution and in the ion exchanger, respectively (atm)
P_e	$d_p u / D_L$, Peclet number
$\overline{P}_e$	$d_p u / \overline{D}$, Peclet number
P_0	standard pressure (atm)
ΔP	pressure drop for a gel bed (atm)
q	parameter in Eq. (8.3) (%)
Q_e	total amount of the protein eluted (g)
Q_s	amount of the protein applied (g)
Q	amount of exchangeable counterions in solution in Eq. (3.37) (mol/g)
$\overline{Q}$	weight capacity of the ion exchanger in Eq. (3.37) (mol/g)
Q_{max}	parameter in Eq. (8.3) (%)
r	distance from the center of the particle (cm)
r_s	equivalent radius of a solute molecule (cm)
r_r	radius of the rod (cm)

R	$1/(1 + HK)$
R	gas constant (atm L/mol K)
R'	$N'_p/(1 + HK')$
Re	$d_p u \rho / \eta$, Reynolds number
R_f	moving velocity of zone mobile-phase velocity
R_p	radius of gel particle (cm)
R_s	resolution defined by Eq. (2.51)
s	empirical constant in Eq. (3.37) (mol/g)
S	$N_p/(1 + HK_R)$
Sh	$k_f d_p / D_m$, Sherwood number
t	time (min or s)
t_R	peak position (retention time) (min or s)
t_0	sample injection time (min or s)
t^*	ut/d_p
T	(absolute) temperature (K or °C)
u	$F/(A\varepsilon) = FZ/V_0$ linear mobile-phase velocity or actual velocity of the solvent flow (cm/min)
u_0	$F/A = FZ/V_t$ superficial velocity of the solvent flow (cm/min)
v	partial molar volume of salt ions in Eq. (3.55) (L mol^{-1})
v_{AB}	partial molar volume of the complex of A and B (L mol^{-1})
v_i	partial molar volume of species i (L mol^{-1})
V_e	elution volume (cm^3)
V_i	gel inside volume (cm^3)

V_G	$Ft_G = V_1 + V_2$ or $V_1/(1 - F_2/F)$, gradient volume (cm^3)
V_0	$AZ\epsilon$ void volume of the column (cm^3)
V_t	total volume of the column (cm^3)
V_t	total volume of a protein solution (solution plus ion exchanger) in Eq. (4.28) (cm^3)
V_1	volume of the mixing vessel of the apparatus for producing gradient (cm^3)
V_2	volume of the reservoir of the apparatus for producing linear gradient (cm^3)
V'	$V_0 + K'(V_t - V_0)$, elution volume of salt (cm^3)
W	width of the elution curve at the baseline (min, mL or dimensionless)
W	amount of the swollen gels (g)
W_s	amount of the protein solution (g)
W_h	peak width measured at $C = C_{max}e^{-1}$ (min)
X	the ratio of the enzyme activity to that of the sample
$X_i, \overline{X}_i$	mole fractions of species i in the outer solution and in the ion exchanger, respectively
X_0	sample volume$/V_0 = t_0/\tau$
y	counterions exchanged per unit weight of ion exchanger in Eq. (3.36) (mol/g)
Y	$[D_m I_a Z/(GHud_p{}^2)]^{1/2}$
z	distance from the inlet of the column (cm)
z_p	distance of the peak position of the protein zone from the inlet of the column (cm)

z_r	distance from the peak position of the protein zone $= z - ut/(1 + HK_R)$ (cm)
Δz	Z/N_p (cm)
z^*	z/Z in Chap. 2
z^*	z/d_p in Chap. 5
Δz^*	$1/N_p$
Z	column length (cm)
Z_i, $\overline{Z}_i$	electrochemical valences of species i in the outer solution and in the ion exchanger, respectively
Z_0	length of the column in which the sample occupies before elution (cm)

Greek Symbols

α	$KW/[V_t(1 - W/\rho_g)\rho_g]$ in Eq. (4.28)
α	real part of Laplace transform variable p (s^{-1})
β	imaginary part of Laplace transform variable p (s^{-1})
γ_i, $\overline{\gamma}_i$	activity coefficient of species i in the outer solution and in the ion exchanger, respectively
γ_m	obstructive tortuosity factor in the intraparticle space or labyrinth factor for the tortuous flow channels
γ_{sm}	$\overline{D}/D_m$
γ_1	$(1/2)\{[1 + (4/3)\alpha]^{1/2} + 1\}$ in Eq. (4.32)
γ_2	$\gamma_1 - 1$ in Eq. (4.33)
Γ_1, Γ_2, Γ_3, Γ_4	defined by Eqs. (3.8), (3.30), (3.31) and (3.32), respectively

δ	$Z/(D_L/u)$
ε	V_0/V_t, void fraction of the column
ε_p	void fraction of the pore
η	viscosity of the elution buffer (cp)
η_i	electrochemical potential of species i (atm L/mol or J/mol)
$\bar{\eta}_i$	η_i in the ion exchanger (atm L/mol or J/mol)
η_i^P, $\bar{\eta}_i^P$	η_i and $\bar{\eta}_i$ at pressure P (atm L/mol or J/mol)
θ	dimensionless time (= $tv/V_0 = tu/Z$)
θ_r	dimensionless time variable in the coordinate system in App. E
θ_R	$t_R/(Z/u)$, θ at peak position of elution curve
λ	packing characterization factor or labyrinth factor for eddy diffusion
λ	defined in Eq. (4.29)
λ_k	kth root of λ in Eq. (4.29)
μ_1	chemical potential of species i (atm L/mol or J/mol)
μ_{i0}	μ_i in the standard state (atm L/mol or J/mol)
μ_i^P	μ_i at pressure P (atm L/mol or J/mol)
μ_i^{P0}	μ_i at standard pressure P_0 (atm L/mol or J/mol)
μ_n'	nth normalized statistical moment of the elution curve (min^n)
μ_n	nth normalized central moment of the elution curve (min^n)
μ_1'	first normalized statistical moment (min)
μ_2	second normalized central moment (min^2)
$\mu_{2,p}$	defined by Eq. (5.4) (min^2)

ν	$d_p u / D_m$, reduced velocity
ξ	$z^* - R\theta$
Π	swelling pressure (atm)
ρ_g	density of the swollen gel (g/cm^3)
ρ_s	density of the solution (g/cm^3)
σ_θ	standard deviation of elution curve with respect to θ
σ_n	standard deviation of elution curve with respect to n
τ	Z/u mean residence time of nonretained solute (min or s)
ϕ	$R_p (u/\bar{D}Z)^{1/2}$
$\phi, \bar{\phi}$	electric potential in the solution and in the ion exchanger, respectively (J/C)
ω	signs of fixed ionic groups in the ion exchanger
ω_i	signs of ionic species i

Subscripts

A	displacement ion
B	counterion
Cl	Cl^- ion
H	H^+ ion
Na	Na^+ ion
NaCl	NaCl
OH	OH^+ ion
P	protein
R	value at peak position, or ion exchanger (or fixed ionic group in the ion exchanger)
RCl	complex of Cl^- and the fixed ionic group

RP	complex of protein and the fixed ionic group
W	water
0	initial value or value at $t = 0$

Superscripts

| — | value in the stationary phase (ion exchanger or resin) |
| ~ | Laplace transform |

Note

The dimension M is used as $mol/l = mol/dm^3$.

References

Adachi, S., Kawamura, Y., Nakanishi, K., and Kamibuko, T. (1978). Agric. Biol. Chem., 42, 1707.

Adams, B. A., and Holmes, E. L. (1935). J. Soc. Chem. Ind., 54, 1.

Altgelt, K. H. (1970). Separ. Sci., 5, 777.

Aranyi, P., and Boross, L. (1974). J. Chromatogr., 89, 239.

Arnold, F. H., and Blanch, H. W. (1986). J. Chromatogr., 355, 13.

Arnold, F. H., Blanch, H. W., and Wilke, C. R. (1985a). J. Chromatogr., 330, 159.

Arnold, F. H., Blanch, H. W., and Wilke, C. R. (1985b). Chem. Eng. J., 30, B9.

Arnold, F. H., Blanch, H. W., and Wilke, C. R. (1985c). Chem. Eng. J., 30, B25.

Ayers, J. S. (1985). Biotechnol. Bioeng., 27, 1721.

Bailey, J. E., and Ollis, D. F. (1986). Biochemical Engineering Fundamentals, 2nd. ed. McGraw-Hill, New York.

Bardsley, W. G., and Wardell, J. M. (1982). J. Chromatogr., 242, 209.

Barker, P. E., and Chauh, C. H. (1981). Chem. Eng., 371/2, 389.

Barker, P. E., and Thawait, S. (1986). Chem. Eng. Res. Des., 64, 302.

Begovich, J. M. and Sisson, W. G. (1984). AIChE J., 30, 705.

Bergenhem, N., Carlsson, U., and Klasson, K. (1985). J. Chromatogr., 319, 59.

Bidlingmeyer, B. A., and Vincent Warren, F. (1984). Anal. Chem., 56, 1583A.

Bjork, W. (1959). J. Chromatogr., 2, 536.

Blanchard, J. S. (1984). Methods Enzymol., 104, 404.

Boardman, N. K., and Partridge, S. M. (1953). Nature, 171, 208.

Boardman, N. K., and Partridge, S. M. (1955). Biochem. J., 59, 543.

Bock, R. M., and Ling, N.-S. (1954). Anal Chem., 26, 1543.

Bogue, D. C. (1960). Anal. Chem., 32, 1777.

Boman, H. G. (1955). Biochim. Biophys. Acta, 16, 245.

Bonnerjae, J., Oh, S., Hoare, M., and Dunnill, P. (1986). Bio/Technology, 4, 955.

Brautigan, D. L., Ferguson-Miller, S., and Margoliash, E. (1978). J. Biol. Chem., 253, 130.

Brewer, S. J. Dickerson, C. H., Ewbank, J., and Fallon, A. (1986). J. Chromatogr., 362, 443.

Broughton, D. B. (1968). Chem. Eng. Prog. 64, No. 8, 60.

Bruton, C., Jakes, R., and Atkinson, T. (1975). Eur. J. Biochem., 59, 327.

Buchholz, K., and Godelmann, B. (1978). Enzyme Engineering, Vol. 4, ed. by G. B. Broun, G. Manecke, and L. B. Wingard. Pergamon Press, New York, p. 89.

Burke, D. J., Duncan, J. K., Dunn, L. C., Cummings, L., Siebert, C. J., and Ott, G. S. (1986). J. Chromatogr., 353, 425.

Burns, M. A., and Graves, D. J. (1985). Biotechnol. Prog., 1, 95.

Buys, T. S., and De Clerk, K. (1972a). J. Chromatogr., 67, 1.

Buys, T. S., and De Clerk, K. (1972b). J. Chromatogr., 67, 13.

Carman, P.-C. (1937). Trans. Inst. Chem. Eng., 15, 150.

Carman, P.-C., and Haul, P. A. (1954). Proc. Roy. Soc., 222-A, 109.

Chang, S. H., Gooding, K. M., and Regnier, F. E. (1976). J. Chromatogr., 120, 321.

Chang, C.-T., McCoy, B. J., and Carbonell, R. G. (1980). Biotechnol. Bioeng., 22, 377.

Charm, S. E., and Matteo, C. C. (1971). Methods Enzymol., 22, 476.

Chase, H. A. (1984a). Chem. Eng. Sci., 39, 1099.

Chase, H. A. (1984b). J. Chromatogr., 297, 179.

Chase, H. A. (1985). Discovery and Isolation of Microbial Products, ed. by M. S. Verrall, Ellis Horwood, England, p. 129.

Chen, H. T., Hsieh, T. K., Lee, H. C., and Hill, F. B. (1977). AIChE J., 23, 695.

Clonis, Y. D., Jones, K., and Lowe, C. R. (1986). J. Chromatogr., 363, 31.

Cluff, J. R., and Hawkes, S. J. (1976). J. Chromatogr. Sci., 14, 248.

Cohn, E. J., Strong, L. E., Hughes, W. L., Mulford, D. J., Ashworth, J. N., Melin, M., and Taylor, H. L. (1946). J. Amer. Chem. Soc., 68, 459.

Cohn, W. E. (1949). Science, 109, 377.

Coq, B., Cretier, G., and Rocca, J. L. (1979). J. Chromatogr., 186, 457.

Crank, J. (1975). The Mathematics of Diffusion, 2nd. ed. Clarendon Press, Oxford.

Crump, K. S. (1976). J. ACM., 23, 89.

Curling, J. M. (ed.) (1980). Methods of Plasma Protein Fractionation. Academic Press, New York.

Curling, J. M., and Cooney, J. M. (1982). J. Parent. Sci. Technol., 36, 59.

DeClerk, K., and Buys, T. S. (1971). J. Chromatogr., 63, 193.

Deisler, P. F., and Wilhelm, R. H. (1953). Ind. Eng. Chem., 45, 1219.

Determann, H., and Brewer, J. E. (1975). In Chromatography, 3rd., ed. by E. Heftmann. Van Nostrand-Reinhold, New York, p. 362.

DeVault, D. (1943). J. Amer. Chem. Soc., 65, 532.

Do, D. D., and Rice, R. G. (1986). AIChE J., 32, 149.

Donnan, F. G. (1911). Z. Elektrochem., 17, 572.

Douzou, P., and Balny, C. (1978). Adv. Prot. Chem., 32, 77.

Drake, B. (1955). Ark. Kem., 8, 1.

Dubner, H., and Abate, J. (1968). J. ACM., 15, 115.

Dunckhorst, F. T., and Houghton, G. (1966). I&EC Fundament., 5, 93.

Dunnill, P., and Lilly, M. D. (1972). Biotechnol. Bioeng. Symp., 3, 97.

Dwyer, J. L. (1984). Bio/Technol., 2, 957.

Edwards, V. H., and Helft, J. M. (1970). J. Chromatogr., 47, 490.

Ekstrom, B., and Jacobson, G. (1984). Anal. Biochem., 142, 134.

Engelhardt, H., and Ahr, G. (1983). J. Chromatogr., 282, 385.

El Rassi, Z., and Horvath, C. (1986). J. Chromatogr., 359, 255.

Fasold, H. (1975). In Chromatography, 3rd. ed., ed. by E. Heftmann. Van Nostrand-Reinhold, New York, p. 466.

Fick, A. (1855). Ann. Phys. Lpz., 170, 59.

Fox, J. B., Calhoun, R. C., and Eglinton, W. J. (1969). J. Chromatogr., 43, 48.

Freiling, E. C., (1955). J. Amer. Chem. Soc., 77, 2067.

Frenz, J., and Horvath, C. (1985). AIChE J., 31, 400.

Frolik, C. A., Dart, L. L., and Sporn, M. B. (1982). Anal. Biochem., 125, 203.

Ganzi, G. C. (1984). Methods Enzymol., 104, 391.

Ghrist, B. F. D., Stadalius, M. A., and Snyder, L. R. (1987). J. Chromatogr., 387, 1.

Gibbs, S. J., and Lightfoot, E. N. (1986). Ind. Eng. Chem. Fundam., 25, 490.

References

381

Giddings, J. C. (1958). J. Chem. Ed., 35, 588.

Giddings, J. C. (1965). Dynamics of Chromatography. Marcel Dekker, New York.

Giddings, J. C. (1967). Anal. Chem., 39, 1027.

Giddings, J. C. (1975). In Chromatography, 3rd. ed., ed. by E. Heftmann. Van Nostrand-Reinhold Company, New York, p. 27.

Giddings, J. C. (1984). Anal. Chem., 56, 1258A.

Giddings, J. C., and Eyring, H. (1955). J. Phys. Chem., 59, 416.

Ginzburg, B. Z., and Cohen, D. (1964). Trans. Faraday Soc., 60, 185.

Gladden, J. K., and Dole, M. (1953). J. Amer. Chem. Soc., 75, 3900.

Glueckauf, E. (1955a). Trans. Faraday Soc., 51, 34.

Glueckauf, E. (1955b). Trans. Faraday Soc., 51, 1540.

Glueckauf, E., and Coates, J. I. (1947). J. Chem. Soc., 1315.

Glueckauf, E., and Duncan, J. F. (1951). Rep., CR. 808, Atomic Energy Research Establishment, Harwell.

Glueckauf, E., and Duncan, J. F. (1952). Rep., CR. 809, Atomic Energy Research Establishment, Harwell.

Glueckauf, E., Barker, K. H., and Kitt, G. P. (1949). Faraday Soc. Discussions, 7, 199.

Gooding, K. M., and Schmuck, M. N. (1984). J. Chromatogr., 296, 321.

Gooding, K. M., and Schmuck, M. N. (1985). J. Chromatogr., 327, 139.

Graham, E. E., and Fook, C. F. (1982). AIChE J., 28, 245.

Gregor, H. P. (1948). J. Amer. Chem. Soc., 70, 1293.

Gregor, H. P. (1951). J. Amer. Chem. Soc., 73, 642.

Grubner, O., and Underhill, D. (1972). J. Chromatogr., 73, 1.

Grushka, E. (1975). In Methods of Protein Separation, Vol. 1, ed. by N. Catsimpoolas. Plenum Press, New York, p. 161.

Grushka, E., Snyder, L. R., and Knox, J. H. (1975). J. Chromatogr. Sci., 13, 25.

Guiochon, G., and Martin, M. (1985). J. Chromatogr., 326, 3.

Gustafsson, J.-G., Frej, A.-K., and Hedman, P. (1986). Biotechnol. Bioeng., 28, 16.

Haarhoff, P. C., and Van Der Linde, H. J. (1966). Anal. Chem., 38, 573.

Haff, L. A., Fagerstam, L. G., and Barry, A. R. (1983). J. Chromatogr., 266, 409.

Hartley, R. W., Peterson, E. A., and Sober, H. A. (1962). Biochemistry, 1, 60.

Hashimoto, K., Adachi, S., Noujima, H., and Maruyama, H. (1983). J. Chem. Eng. Japan, 16, 400.

Hawkes, S. J. (1972). J. Chromatogr., 68, 1.

Hearn, M. T. W. (1984). Methods Enzymol., 104, 190.

Hearn, M. T. W., Regnier, F. E., and Wehr, C. T. (eds.) (1983). High-Performance Liquid Chromatography of Proteins and Peptides. Academic Press, New York.

Heftmann, E. (ed.) (1975). Chromatography, 3rd. ed. Van Nostrand-Reinhold Company, New York.

Helfferich, F. (1959). Ionenaustausher, Verlag Chemie GmbH.

Helfferich, F. (1959). Ion Exchange, McGraw-Hill, New York.

Helfferich, F., and James, D. B. (1970). J. Chromatogr., 46, 1.

Henrion, P. N. (1964). Trans. Faraday Soc., 60, 72.

Hills, J. H. (1986). Chem. Eng. Sci., 41, 2779.

Hirs, C. H. W., Stein, W. H., and Moore, S. (1951). J. Amer. Chem. Soc., 73, 1893.

Hjerten, S. (1962). Arch. Biochem. Biophys., 99, 466.

Hjerten, S., and Mosbach, R. (1962). Anal. Biochem., 3, 109.

Hodgman, C. D. (ed.) (1966). Handbook of Biochemistry and Physics, 47th ed. Chemical Rubber Publishing, Cleveland, p. 2274.

Hofstee, B. H. J. (1976). In Methods of Protein Separation, Vol. 2, ed. by N. Catsimpoolas. Plenum Press, New York, p. 245.

Hollein, H. C., Ma, H.-C., Huang, C.-R., and Chen, H. T. (1982). Ind. Eng. Chem. Fundam., 21, 205.

Horowitz, S. B., and Fenichel, I. R. (1964). J. Phys. Chem.,
68, 3378.

Horvath, C., and Lin, H.-J. (1976). J. Chromatogr., 126,
401.

Horvath, C., and Lin, H.-J. (1978). J. Chromatogr., 149, 43.

Houghton, G. (1963). J. Phys. Chem., 67, 84

Houghton, G. (1964). J. Chromatogr., 15, 5.

Huang, J.-C., Rothstein, D., and Madey, R. (1983). J.
Chromatogr., 261, 1.

Huang, S. Y., Lin, C. K., Chang, W. H., and Lee, W. S.
(1986). Chem. Eng. Commun., 45, 291.

Huisman, T. H. J., and Dozy, A. M. (1965). J. Chromatogr.,
19, 160.

Hupe, K. P., Jonker, R. J., and Rozing, G. (1984). J.
Chromatogr., 285, 253.

Ikigai, H., Nakae, T., and Kato, Y. (1985). J. Chromatogr.,
322, 212.

International Critical Tables (1929). Vol. V. McGraw-Hill,
New York, p. 71.

James, K., and Stanworth, D. R. (1964). J. Chromatogr., 15,
324.

Janatova, J., Fuller, J. K., and Hunter, M. J. (1968). J.
Biol. Chem., 243, 3612.

Jandera, P., and Churacek, J. (1974a). J. Chromatogr., 91,
223.

Jandera, P., and Churacek, J. (1974b). J. Chromatogr., 93,
17.

Janson, J.-C. (1971). J. Agr. Food. Chem., 19, 581.

Janson, J.-C., and Hedman, P. (1982). Advances in Biochemical Engineering, Vol. 25, ed. by A. Fiechter. Springer-
Verlag, New York, p. 43.

Janson, J.-C., and Hedman, P. (1987). Biotechnol. Prog., 3,
9.

Johansson, B.-L., and Stafstrom, N. (1984). J. Chromatogr.,
314, 396.

Johnson, T. J. A., and Bock, R. M. (1974). Anal. Biochem., 59, 375.

Josic, D., Hofmann, W., and Reutter, W. (1986). J. Chromatogr., 371, 43.

Kadoya, T., Isobe, T., Amano, Y., Kato, Y., Nakamura, K., Okuyama, T. (1985). J. Liquid Chromatogr., 8, 635.

Kaminski, M., Klawiter, J., and Kowalczyk, J. S. (1982). J. Chromatogr., 243, 225.

Kaminski, M., Kandybowicz, B., and Kowalczyk, J. S. (1984). J. Chromatogr., 292, 85.

Kato, Y., and Hashimoto, T. (1985c). HRC & CC, 8, 78.

Kato, Y., Komiya, K., Iwaeda, T., Sasaki, H., and Hashimoto, T. (1981a). J. Chromatogr., 205, 185.

Kato, Y., Komiya, K., Iwaeda, T., Sasaki, H., and Hashimoto, T. (1981b). J. Chromatogr., 206, 135.

Kato, Y., Nakamura, K., and Hashimoto, T. (1982a). J. Chromatogr., 245, 193.

Kato, Y., Komiya, K., and Hashimoto, T. (1982b). J. Chromatogr., 246, 13.

Kato, Y., Nakamura, K., and Hashimoto, T. (1982c). J. Chromatogr., 253, 219.

Kato, Y., Nakamura, K., and Hashimoto, T. (1983a). J. Chromatogr., 256, 143.

Kato, Y., Nakamura, K., and Hashimoto, T. (1983b). J. Chromatogr., 266, 385.

Kato, Y., Nakamura, K., Yamazaki, Y., and Hashimoto, T. (1985a). J. Chromatogr., 318, 358.

Kato, Y., Nakamura, K., and Hashimoto, T. (1985b). HRC & CC, 8, 154.

Kato, Y., Kitamura, T., Mitsui, A., and Hashimoto, T. (1987). J. Chromatogr., 398, 327.

Katoh, S., and Sada, E. (1980a). J. Chem. Eng. Japan, 13, 151.

Katoh, S., and Sada, E. (1980b). J. Chem. Eng. Japan, 13, 336.

Katz, E., Ogan, K. L., and Scott, R. P. W. (1983). J. Chromatogr., 270, 51.

Kawasaki, T., and Bernardi, G. (1970). Biopolymers, 9, 257, 269.

Kelly, R. N., and Billmeyer, F. W. (1969). Anal. Chem., 41, 874.

Kennedy, G. J., and Knox, J. H. (1972). J. Chromatogr. Sci., 10, 549.

Kirkegaard, L. H. (1973). Biochemistry, 12, 3627.

Kirkegaard, L. H. (1976). Methods of Protein Separation, Vol. 2, ed. by N. Catsimpoolas. Plenum Press, New York, p. 279.

Kirkegaard, L. H., Johnson, T. J. A., and Bock, R. M. (1972). Anal. Biochem., 50, 122.

Klawiter, J., Kaminski, M., and Kowalczyk, J. S. (1982). J. Chromatogr., 243, 207.

Knight, C. S., Weaver, V. C., and Brook, B. N. (1963). Nature, 200, 245.

Knox, J. H., and Pyper, H. M. (1986). J. Chromatogr., 363, 1.

Knox, J. H., and Scott, H. P. (1983). J. Chromatogr., 282, 297.

Kopaciewicz, W., and Regnier, F. E. (1983a). Anal. Biochem., 129, 472.
Kopaciewicz, W., and Regnier, F. E. (1983b). Anal. Biochem., 133, 251.

Kopaciewicz, W., Rounds, M. A., Fausnaugh, J., and Regnier, F. E. (1983). J. Chromatogr., 266, 3.

Kopaciewicz, W., Rounds, M. A., and Regnier, F. E. (1985). J. Chromatogr., 318, 157.

Krstulovic, A. M., and Brown, P. R. (1982). Reversed-Phase High-Performance Liquid Chromatography. John Wiley & Sons, New York.

Kubin, M. (1965). Collect. Czech. Chem. Commun., 30, 1104.

Kucera, E. (1965). J. Chromatogr., 19, 237.

Kuntz, I. D., and Kauzmann, W. (1974). Adv. Prot. Chem.,
28, 239.

Lampson, G. P., and Tytell, A. A. (1965). Anal. Biochem.,
11, 374.

Langer, G., Roethe, A., Roethe, K. P., and Gelbin, D. (1978).
Int. J. Heat Mass Transfer, 21, 751.

Lapidus, L. (1962). Digital Computation for Chemical Engineers. McGraw-Hill, New York, p. 89.

Lapidus, L., and Amundson, N. R. (1952). J. Phys. Chem.,
56, 984.

Larre, C., and Gueguen, J. (1986). J. Chromatogr., 361,
169.

Lazare, L., Sundheim, B. R., and Gregor, H. P. (1956).
J. Phys. Chem., 60, 641.

Leaback, D. H., and Robinson, H. K. (1975). Biochem. Biophys. Res. Commun., 67, 248.

Lehninger, A. L. (1970). Biochemistry. Worth Publishers,
New York.

LeRoith, D., Shiloach, J., and Leahy, T. J. (eds.) (1985).
Purification of Fermentation Products. ACS Symp. Ser., 271,
113.

Lewis, W. K., and Whitman, W. G. (1924). Ind. Eng. Chem.,
16, 1215.

Lightfoot, E. N., Sanchez-Palma, R. J., and Edwards, D. O.
(1962). H. M. Schoen (ed.), New Chemical Engineering
Separation Technique, Interscience, New York, p. 99.

Luyben, K. C. A. M., Liou, J. K., and Bruin, S. (1982).
Biotechnol. Bioeng., 24, 533.

Majors, R. E. (1977). J. Chromatogr. Sci., 15, 334.

Martin, A. J. P., and Synge, R. L. M. (1941). Biochem. J.,
35, 1358.

Mayer, S. W., and Tompkins, E. R. (1947). J. Amer. Chem.
Soc., 69, 2866.

McCoy, B. J. (1985a). Ind. Eng. Chem. Fundam., 24, 500.

McCoy, B. J. (1985b). Biotechnol. Bioeng., 27, 1477.

McCoy, B. J. (1986). AIChE J., 32, 1570.

Mehta, R. V., Merson, R. L., and McCoy, B. J. (1973).
AIChE J., 19, 1068.

Miller, R. R., Peters, S. P., Kuhlenschmidt, M. S., and Glew,
R. H. (1976). Anal. Biochem., 72, 45.

Moore, J. C. (1970). Separ. Sci., 5, 723.

Moore, S., and Stein, W. H. (1951). J. Biol. Chem., 192, 663.

Morris, C. J. O. R., and Morris, P. (1964). Separation Methods in Biochemistry, 2nd ed. Pitman, London.

Nakamura, K., and Kato, Y. (1985). J. Chromatogr., 333, 29.

Nakanishi, K., Adachi, S., Yamamoto, S., Matsuno, R.,
Tanaka, A., and Kamikubo, T. (1977a). Agric. Biol. Chem.,
41, 2455.

Nakanishi, K., Yamamoto, S., Matsuno, R., and Kamikubo, T.
(1977b). Agric. Biol. Chem., 41, 1465.

Nakanishi, K., Yamamoto, S., Matsuno, R., and Kamikubo, T.
(1978). Agric. Biol. Chem., 42, 1943.

Nakanishi, K., Yamamoto, S., Matsuno, R., and Kamikubo, T.
(1979). Agric. Biol. Chem., 43, 2507.

Nakanishi, K., Matsuno, R., and Kamikubo, T. (1983). Memoirs of the College of Agriculture, Kyoto University, No. 120,
19.

Namikawa, R., Okazaki, H., Nakanishi, K., Matsuno, R., and
Kamikubo, T. (1977). Agric. Biol. Chem., 41, 1003.

Nelson, W. C., Silarski, D. F., and Wankat, P. C. (1978).
Ind. Eng. Chem. Fundam., 17, 32.

Nernst, W. (1888). Z. Phys. Chem., 2, 613.

Nernst, W. (1889). Z. Phys. Chem., 4, 129.

Nishimoto, K., Nakahara, Y., Matsubara, K., and Sakamoto, K.
(1987). Toyo Soda Kenkyu Hokoku, 31, 45.

Novotny, J. (1971). FEBS Lett., 14, 7.

Novotny, M. (1981). Anal. Chem., 53, 1294A.

Ogston, A. G. (1958). Trans. Faraday Soc., 54, 1754.

Parente, E. S., and Wetlaufer, D. B. (1986). J. Chromatogr.,
355, 29.

Peacock, R. (1986). Prog. Biol. Technol., 4, 5.

Perrin, D. D., and Dempsey, B. (1974). Buffers for pH and Metal Ion Control. Chapman and Hall, London.

Peterson, E. A. (1970). Cellulosic ion exchangers. In Laboratory Techniques in Biochemistry and Molecular Biology, Vol. 2, ed. by T. S. Work and E. Work. North Holland, Amsterdam, p. 228.

Peterson, E. A. (1978). Anal. Biochem., 90, 767.

Peterson, E. A., and Sober, H. A. (1956). J. Amer. Chem. Soc., 78, 751.

Peterson, E. A., and Torres, A. R. (1983). Anal. Biochem., 130, 271.

Peterson, E. A., and Torres, A. R. (1984). Methods Enzymol., 104, 113.

Pharmacia Fine Chemicals (1987). Ion Exchange Chromatography: Principles and Methods. Pharmacia Fine Chemicals, Sweden.

Pitt, W. W. (1976). J. Chromatogr., Sci., 14, 396.

Planck, M. (1890). Ann. Phys., 39, 161.

Poppe, H., and Kraak, J. C. (1983). J. Chromatogr., 255, 395.

Porath, J. (1972). Biotechnol. Bioeng. Symp., 3, 145.

Porath, J., and Flodin, P. (1959). Nature, 183, 1657.

Porath, J., Janson, J. C., and Laas, T. (1971). J. Chromatogr., 60, 167.

Raghavan, N. S., and Ruthven, D. M. (1983). AIChE J., 29, 922.

Ramachandran, P. A., and Smith, J. M. (1978). Ind. Eng. Chem. Fundam., 17, 148.

Rasmuson, A., and Neretnieks, I. (1980). AIChE J., 26, 686.

Regnier, F. E. (1983). Methods Enzymol., 91, 137.

Regnier, F. E. (1984). Methods Enzymol., 104, 170.

Regnier, F. E., and Gooding, K. M. (1980). Anal. Biochem., 103, 1

Regnier, F. E., and Mazsaroff, I. (1987). Biotechnology Progress, 3, 22.

Reilley, C. N., Hildebrand, G. P., and Ashley, J. W. (1962). Anal. Chem., 34, 1198.

Rhee, H.-K., and Amundson, N. R. (1982). AIChE J., 28, 423.

Rhodes, M. B., Azari, P. R., and Feeney, R. E. (1958). J. Biol. Chem., 230, 399.

Richey, J. S. (1984). Methods Enzymol., 104, 223.

Righetti, P. G., and Caravaggio, T. (1976). J. Chromatogr., 127, 1.

Righetti, P. G., Tudor, G., and Ek, K. (1981). J. Chromatogr., 220, 115.

Robinson, R. A., and Stokes, R. H. (1955). Electrolyte Solutions. Butterworths, London.

Ruckenstein, E., and Lesins, V. (1986). Biotechnol. Bioeng., 28, 432.

Sada, E., Katoh, S., Inoue, T., Matsukura, K., Shiozawa, M., and Takeda, A. (1984). J. Chem. Eng. Japan, 17, 642.

Sada, E., Katoh, S., Inoue, T., and Shiozawa, M. (1985). Biotechnol. Bioeng., 27, 514.

Sada, E., Katoh, S., Kondo, Λ., and Ishida, S. (1987). J. Ferment. Technol., 65, 111.

Sada, E., Katoh, S., and Shiozawa, M. (1982). Biotech. Bioeng., 24, 2279.

Said, A. S. (1956). AIChE J., 2, 477.

Saint-Blancard, J., Kirzin, J. M., Riberon, P., Petit, F., Fourcart, J., Girot, P., and Boschetti, E. (1982). In Affinity Chromatography and Related Techniques, ed. by T. C. J. Gribnau, J. Visser, and R. J. F. Nivard. Elsevier, Amsterdam.

Saunders, D. L. (1975). In Chromatography, 3rd. ed., ed. by E. Heftmann. Van Nostrand-Reinhold, New York, p. 77.

Schwab, H., Rieman, W., and Vaughan, P. A. (1957). Anal. Chem., 29, 1357.

Schwartz, C. E., and Smith, J. M. (1953). Ind. Eng. Chem., 45, 1209.

Scopes, R. K. (1977a). Biochem. J., 161, 253.

Scopes, R. K. (1977b). Biochem. J., 161, 265.

Scopes, R. K. (1982). Protein Purification, Springer-Verlag, New York.

Scott, C. D., Pitt, W. W., and Johnson, W. F. (1974). J. Chromatogr., 99, 35.

Scott, R. P. W. (1980). J. Chromatogr. Sci., 18, 49.

Scott, R. W., Duffy, S. A., Moellering, B. J., and Prior, C. (1987). Biotechnol. Prog., 3, 49.

Sherwood, T. K., Pigford, R. L., and Wilke, C. R. (1975). Mass Transfer, McGraw-Hill, New York.

Siegel, J. H., Dupre, G. D., Pirkle, J. C. (1986). Chem. Eng. Prog., 82, 57.

Sluyterman, L. A. A.-E., and Elgersma, O. (1978). J. Chromatogr., 150, 17.

Sluyterman, L. A. A.-E., and Wijdenes, J. (1978). J. Chromatogr., 150, 31.

Smith, M. H. (1968). Handbook of Biochemistry, 2nd ed., ed. by H. A. Sober. CRC Press, Cleveland, C3.

Snyder, L. R., and Kirkland, J. J. (1974). Introduction to Modern Liquid Chromatography. John Wiley & Sons, New York.

Snyder, L. R., and Saunders, D. L. (1969). J. Chromatogr. Sci., 7, 195.

Snyder, L. R., Dolan, J. W., and Gant, J. R. (1979). J. Chromatogr., 165, 3.

Snyder, L. R., Stadalius, M. A., and Quarry, M. A. (1983). Anal. Chem., 55, 1413A.

Sober, H. A., and Peterson, E. A. (1958). Fed. Proc., 17, 1116

Stadalius, M. A., Gold, H. S., and Snyder, L. R. (1984). J. Chromatogr., 296, 31.

Stadalius, M. A., Ghrist, B. F. D., and Snyder, L. R. (1987). J. Chromatogr., 387, 21.

Stampe, D., Wieland, B., and Kohle, A. (1986). J. Chromatogr., 363, 101.

Stone, M. G. T. (1962). Trans. Faraday Soc., 58, 805.

Stout, R. W., Sivakoff, S. I., Ricker, R. D., and Snyder, L. R. (1986). J. Chromatogr., 353, 439.

Strobel, G. J. (1982). Chem. Technol., 11, 1354.

Suzuki, M. (1974). J. Chem. Eng. Japan, 7, 262.

Suzuki, M., and Smith, J. M. (1972). Chem. Eng. J. 3, 256.

Svensson, H. (1961). Acta Chem. Scand., 15, 325.

Tandy, N. E., Dilly, R., and Regnier, F. E. (1983). J. Chromatogr., 266, 599.

Thompson, H. M. (1850). J. Roy. Agric. Soc. Engl., 11, 68.

Tiselius, A. (1943). Ark. Kemi. Min. Geol., 16A, 18.

Torres, A. R., Dunn, B. E., Edberg, S. C., and Peterson, E. A., (1984). J. Chromatogr., 316, 125.

Torres, A. R., Edberg, S. C., and Peterson, E. A. (1987). J. Chromatogr., 389, 177.

Tsou, H. S., and Graham, E. E. (1985). AIChE J., 31, 1959.

Unger, K. K., and Janzen, R. (1986). J. Chromatogr., 373, 227.

Vageler, P., and Woltersdorf, J. (1930a). Z. Pflanzenernahr. Dung. Bodenkunde, A15, 329.

Vageler, P., and Woltersdorf, J. (1930b). Z. Pflanzenernahr. Dung. Bodenkunde, A16, 184.

van Deemter, J. J., Zuiderweg, F. J., and Klinkenberg, A. (1956). Chem. Eng. Sci., 5, 271.

van der Zee, R., and Welling, G. W. (1984). J. Chromatogr., 292, 412.

Vanecek, G., and Regnier, F. E. (1980). Anal. Biochem., 109, 345.

Vardanis, A. (1985). J. Chromatogr., 350, 299.

Verrall, M. S. (ed.) (1985). Discovery and Isolation of Microbial Products, Ellis Horwood, Chichester, England.

Vermulen, T., LeVan, M. D., Hiester, N. K., and Klein, G. (1984). In R. H. Perry, D. W. Green and J. O. Maloney (eds.), "Perry's Chemical Engineer's Handbook," 6th ed., McGraw-Hill, New York, p. 16—1.

Wakao, N., and Tanaka, K. (1973). J. Chem. Eng. Japan, 6, 338.

Wankat, P. C. (1974a). Anal. Chem., 46, 1400.

Wankat, P. C. (1974b). Separ. Sci., 9, 85.

Wankat, P. C. (1977a). AIChE J., 23, 859.

Wankat, P. C. (1977b). Ind. Eng. Chem. Fundam., 16, 468.

Wankat, P. C. (1984). Ind. Eng. Chem. Fundam., 23, 256.

Way, J. T. (1850). J. Roy. Agric. Soc. Engl., 11, 68, 313.

Wehr, C. T. (1984). Methods Enzymol., 104, 133.

Weiss, J. (1943). J. Chem. Soc., 297.

Wiegner, G., and Jenny, H. (1927). Kolloid Z., 42, 268.

Wilson, E. J., and Geankoplis, C. J. (1966). Ind. Eng. Chem. Fundam., 5, 9.

Wilson, J. N. (1940). J. Amer. Chem. Soc., 62, 1583.

Wood, R., Cummings, L., and Jupille, T. (1980). J. Chromatogr. Sci., 18, 551.

Yamamoto, S., Nakanishi, K., Matsuno, R., and Kamikubo, T. (1978). Agric. Biol. Chem., 42, 963.

Yamamoto, S., Nakanishi, K., Matsuno, R., and Kamikubo, T. (1979). Agric. Biol. Chem., 43, 2499.

Yamamoto, S., Nakanishi, K., Matsuno, R., and Kamikubo, T. (1983a). Biotechnol. Bioeng., 25, 1465.

Yamamoto, S., Nakanishi, K., Matsuno, R., and Kamikubo, T. (1983b). Biotechnol. Bioeng., 25, 1373.

Yamamoto, S., Agawa, M., Nakano, H., and Sano, Y. (1985). In Drying '85, ed., by R. Toei and A. S. Mujumdar. Hemisphere, Washington, D.C., p. 330.

Yamamoto, S., Nomura, M., and Sano, Y. (1986a). J. Chem. Eng. Japan, 19, 227.

Yamamoto, S., Nomura, M., and Sano, Y. (1986b). Paper presented at World Congress III of Chemical Engineering (Tokyo), W-0630.

Yamamoto, S., Nomura, M., and Sano, Y. (1987a). J. Chromatogr., 394, 363.

Yamamoto, S., Nomura, M., and Sano, Y. (1987b). J. Chromatogr., 396, 355.

Yamamoto, S., Nomura, M., and Sano, Y. (1987c). J. Chromatography, 409, 101.

Yamamoto, S., Nomura, M., and Sano, Y. (1987d). AIChE J., 33, 1426.

Yamamoto, S., Nomura, M., and Sano, Y. (1987e). Paper presented at 20th Autumn meeting of the Society of Chemical Engineers (Japan), SA314.

Yamamoto, S., Nomura, M., and Sano, Y. (1987f). In preparation.

Yang, C., and Tsao, G. T. (1982). Advances in Biochemical Engineering, Vol. 25, ed. by A. Fiechter. Springer-Verlag, New York, p. 1.

Yang, F. J. (1983). J. High. Res. Chromatogr. Chromatogr. Commun., 6, 348.

Yang, V. C., Bernstein, H., and Langer, R. (1987). Biotechnol. Prog., 3, 27.

Yon, R. J. (1980). Biochem. J., 185, 211.

Young, J. L., and Webb, B. A. (1978). Anal. Biochem., 88, 619.

Young, J. L., Webb, B. A., Coutie, D. G., Reid, B. (1978). Biochem. Soc. Trans., 6, 1051.

Young, M. E., Carroad, P. A., and Bell, R. L. (1980). Biotechnol. Bioeng., 22, 947.

Index